QK661 .M37
Martin, Alexander Campbell,
identification manual
3 4369 00254586 1
W9-AGM-865

WITHDRAWN

Seed Identification Manual

Seed Identification Manual

Alexander C. Martin and William D. Barkley

University of California Press
Berkeley, Los Angeles and London

UNIVERSITY OF CALIFORNIA PRESS, BERKELEY AND LOS ANGELES
UNIVERSITY OF CALIFORNIA PRESS, LTD., LONDON, ENGLAND

SECOND PRINTING, 1973
ISBN: 0-520-00814-6
LIBRARY OF CONGRESS CATALOG CARD NUMBER: 61-7528

Designed by Howard H. Bezanson
MANUFACTURED IN THE UNITED STATES OF AMERICA

Acknowledgments

The Fish and Wildlife Service of the United States Department of the Interior, particularly as represented by its Branch of Wildlife Research, deserves credit for making this manual possible. It contracted for the extensive series of photographs, made available the outstanding seed collection at its Patuxent Wildlife Research Center, Laurel, Maryland, and aided substantially in other ways.

Special acknowledgment is also due Arnold L. Nelson, formerly Director of the Patuxent Wildlife Research Center. His name might well have been included in the authorship because of valuable consultation given during all phases of the work and because his active interest brought essential financial support.

Thanks are also given to many individuals who helped with various aspects of the undertaking.

ALEXANDER C. MARTIN
WILLIAM D. BARKLEY

Contents

Introduction

This manual is an attempt to meet the long-standing need for a reference work dealing exclusively with seed identification.

The ability to recognize seeds, always important in farming, has become even more essential in modern scientific agriculture: without it, there would be little merit in perfecting methods of growing useful plants. But utility for human consumption, whether as food or in the clothing and construction industries, is not the sole criterion for deciding which plants are useful. Increasingly in recent decades, new knowledge about the interdependence of forms of plant and animal life has been made available, and the public has been alerted to the dangers, both aesthetic and economic, of allowing our natural resources to dwindle. Plants formerly thought to be of little worth have been found to be good, because they provide food and cover for wildlife, or protect watersheds by their quick regrowth over damaged lands. The number of plants regarded as useful and the importance of knowing their seeds have grown accordingly.

Of no less importance, however, is the identification of undesirable seeds. On behalf of the crop harvesters and the conservationists, whose ideas of "undesirable" differ somewhat according to their aims, a continual war is being waged against plants that are unwanted for one reason or another. Plant-quarantine officers and chemical manufacturers, agricultural experiment stations and seed-testing laboratories work vigilantly in this unceasing fight, in which millions of dollars are expended every year, and millions more are lost through reduced crop potentials.

While the value of a wide acquaintance with seeds, both "good" and "bad," has thus been growing at an accelerating rate, the tools for obtaining this knowledge have remained grossly inadequate. The seeds of some genera have apparently not been illustrated or described in published form—or at least, not amply enough to identify them. For other genera, the seed characteristics have been reported, but in papers that are not easily accessible. For the most part, with a few notable exceptions, the available information about seeds is given only in bits, here and there, in works mainly concerned with other aspects of plant life. To remedy this situation, the authors of this manual have brought together, for direct observation and study on a comparative basis, pictures and practical descriptions of as large and representative a collection of seeds as possible, within the spatial and other limitations mentioned below under Basis for Inclusion.

The immediate aim of the manual is to help agriculturists, foresters, wildlife biologists, and others interested in land-use programs to identify the seeds in their particular ecological fields of interest. With this practical aim in view, the authors have, in the main, restricted the content of the descriptions to the characteristics useful for identification. Also, bearing in mind the different backgrounds of the probable readers of this manual in respect to scientific preparation and experience, the authors have thought it best to keep the descriptions nontechnical, so far as was feasible, and thus adapt the material to a broad range of interests and skills.

Definition

The term *seed* must here be understood in its broad, popular sense. It is applied not only to true seeds, but also to equivalent structures which look like and function as seeds. Technical designations for seedlike dry fruits are indicated parenthetically in the text.

Plan of the Book

Because the quickest and best way to form a true concept of an object, short of seeing the thing itself, is to examine a good picture of it, the dominant principle governing the order, proportions, and mode of presentation of the contents of this book has been emphasis on illustrations.

The manual consists of two major parts, one wholly photographic except for legends and a few explanatory remarks, and the other primarily textual. The pictorial part is given priority by being the larger and by being placed first. The photographs, most of which are magnified, usually show several views of each seed in detail. The superior effectiveness of pictures suggested setting life-sized silhouettes within the plates themselves, so that the reader could receive an image of the seed's real size at once, and need not attempt the difficult feat of translating numerical magnifications into spatial terms. Again in accord with the visual principle, the textual part of the manual is amply illustrated with

288 figures in the form of line drawings and diagrams, and the generic descriptions are accompanied by references to the corresponding plates.

In the first part of the book, headed Seed Photographs, the photographs are organized systematically within eight plant categories, grouped in three ecological sections with the headings of Farmlands, Wetlands, and Woodlands.

The sectional groupings should simplify and expedite identification by enabling the investigator first to classify a seed roughly by its habitat, thus eliminating the need to compare it with seeds from other habitats. The names of these ecological sections and the designations of general types of plants (Weeds, Crops, Aquatics, and so on) must be recognized as merely convenient, short labels that require some qualification or interpretation, since a certain amount of overlapping in their meaning and application is inevitable. The heading Farmlands is intended in a loose sense, to include not only the fields tilled for crops but also rangelands and other open, uncultivated fields. That is why so many wild plants that are not clearly pest species are included under the subheading Weeds. Conversely, under Wetlands are some grasses which flourish in seasonally damp fields and which therefore sometimes interfere with crops. The heading Woodlands also needs a liberal interpretation. Some of the plants included can grow in open places as well as in wooded areas.

The word *weeds*, in its most derogatory sense, is of course a relative term, its application partly dependent on the objectives of the person who uses it. In this manual the subheading Weeds is interpreted in a broad sense, so as to include in the classification, alongside the plants that are objectionably aggressive or even poisonous, not only many harmless wild plants but also a number that are a valuable source of food and cover for many kinds of wildlife, and some that are very useful as soil binders.

The plant-family sequence used within the limits of the ecological units of Farmlands, Wetlands, and Woodlands and their plant categories has been supplemented by a physical-feature arrangement in one gallery. This will enable investigators, whether botanists or nonbotanists, to identify seeds by simply matching unknowns with pictures which are segregated on the basis of distinctive shapes and appendages. The plan provides an effective short cut, as compared to scanning all or many of the pictures in a family-order gallery, but the additional space it requires can be justified only by an important, large group.

Such a group is that of weed seeds. Because of its special interest to technicians and investigators in several land-management fields, as well as to farmers, plant-quarantine officers, and others, the artificial system of photograph organization has been applied to it. In the subdivision called Weeds, Arranged by Physical Features, winged seeds are grouped together, as are seeds with pappus, others with awns, and still others of special shapes such as spheroid, flattish, or angular. Seeds on the border line between two or more types of shape are often difficult to place satisfactorily, but the preliminary key, figure 1, helps to define the meaning of the designations used and also indicates the series of plates in which each class of related physical features is illustrated. This subdivision, of course, duplicates much of the systematic subdivision on Weeds, though not all of it. Altogether, there are 824 photographic plates, showing the seeds of more than 600 plant species from various parts of the United States.

The common plant names used throughout the manual are based in general on those in Standardized Plant Names, by Harlan P. Kelsey and William A. Dayton, prepared for the American Joint Committee on Horticultural Nomenclature (2d ed.; Harrisburg, Pa., J. Horace McFarland Co., 1942), but some stylistic modifications have been made in the longer and more unwieldy compounds.

The descriptive section on Identification Clues is the second major part of the book. The seeds presented in the photographic guide are here approached analytically. By means of text and figures this part of the manual points out, as its heading implies, the seed characteristics that are especially valuable for identification. It consists of a single series of descriptions with appropriate illustrations, arranged in systematic order. As explained under Procedure in Seed Identification, below, it was thought best to place the emphasis on the seed traits of families and genera; usually, species are mentioned or illustrated only as examples of the larger groups.

Normally, there is a brief treatment of the family traits of the seeds, followed by several seed descriptions for the genera of that family shown in the photographs, two or three genera with very similar seeds occasionally being treated together; and if only one genus of a family is represented the family description is omitted. One marked departure has been made from the rule of describing the seeds genus by genus. The immense and important family of the Gramineae, though represented in the plates by more than forty genera and nearly twice as many species, is discussed chiefly at the family and tribal levels and not in any separate generic descriptions, because the Department of Agriculture's thorough and well-illustrated Manual of Grasses is readily available. Thus it was possible to avoid the numerous and repetitive descriptions without leaving a serious gap.

The section on Identification Clues is intended primarily to enable the reader to confirm the preliminary identifications of families and genera which he has made by examining the photographs. In addition, the information on the generic and family traits of seeds will be serviceable in guiding him to the proper systematic starting point; and in this connection the clues on internal morphology should prove especially helpful.

Basis for Inclusion

It can be anticipated that some will wonder, regretfully, why the manual was limited to 600 species and why this or that particular seed was not included. The reason for this is essentially the same one that necessitated reproducing the photographs in black and white rather than in color, namely, prohibitive cost. The problem of limiting the number of genera and species to a total that could be treated feasibly in one moderate-sized volume was further complicated by the necessity of keeping the selection representative of botanical groups, geographic areas, and the interests of various scientific and economic fields.

A nationwide perspective has been used in choosing genera and species for inclusion in the ecological sections and plant categories. Selection was based, for the most part, on the abundance, widespread distribution, and economic or other importance of the plants or plant groups. Naturally, in some instances the limits for inclusion and exclusion were very difficult to decide. The category Crops was purposely restricted, partly because the public is already well acquainted with most of the more common seeds, and also because there have been many publications on the numerous varieties of cultivated plants.

Procedure in Seed Identification

Sometimes one clue, if extreme and unique, serves alone to identify a seed. But more commonly, several different features in combination are required for a positive determination. Frequently a seed can be identified to the species as reliably as can the whole plant from which it was obtained. More often, however, this is not possible, and in many genera the determinations of identity should be left at the genus level to avoid an unjustifiable danger of error.

The most useful clues for recognition of seeds are usually the shape, the size, and peculiarities of the surface and coloring. Seed surfaces vary from smooth and glossy to dull or rough; surface irregularities include pittings, grooves, and other types of sculpturing. Characteristics of the attachment scar (hilum), particularly its shape, size, and position, are often significant in narrowing the range of genera or species to be considered. Also, the presence of any noticeable external features such as wings, pappus, spines, awns, or hairs is likely to be helpful in placing the seed correctly.

Especially useful clues to the family or genus of an unknown seed are available for those who are willing to delve into its internal characteristics. These include the embryo's shape, size, and position and its relation to endosperm, if the latter is present, and also the wall's thickness, texture, and inner-surface markings. A quick check on a seed's internal morphology can often indicate its family or genus and thus help the investigator avoid serious mistakes about seeds which look surprisingly alike from the outside, but internally are very distinct.

Seed Photographs

This pictorial part of the manual, containing 824 plates, is divided into the three habitat classifications of Farmlands, Wetlands, and Woodlands, each of which is again divided into two or more categories on the basis of general types of plants. Within each plant category the seeds are presented in systematic order; but, in addition, about two-thirds of those in the first and largest category, Weeds, are presented a second time, arranged to exemplify the more important traits of their appearance useful for identification. More than six hundred species from the continental United States are shown.

The seeds pictured here are described in the second part of the manual, headed Identification Clues, in which the discussion of each genus or group of similar genera is keyed to the corresponding plate or plates by cross reference. Reference from plate to text can be made through the Index.

Since nearly all of the seed photographs are magnified, an accurate impression of the size of each seed is conveyed by a small, life-sized photograph in silhouette, set in the lower left corner of the plate; but for the very few seeds too large for the silhouette method (see pls. 553, 555, and 752–755), the inset box is used for indicating size by figures, in the conventional way. Actual measurements are regularly included in the generic descriptions in Identification Clues.

Farmlands

Weeds, Arranged Systematically
(Weeds I)

Farmlands, as construed here, include western rangelands, fields, fencerows, gardens, roadsides, and open places that harbor weedy or other wild plants. Some of the plants in this section are not weeds in the strict sense. Many of the 307 species illustrated are important both to agriculturists and to wildlife biologists. Representatives of the genera shown here are also included in the following unit on Weeds, Arranged by Physical Features.

1. *Bromus secalinus*
Chess

2. *Bromus catharticus*
Rescuegrass

3. *Bromus tectorum*
Cheatgrass

4. *Festuca octoflora*
Fescue

5. *Festuca obtusa*
Fescue

6. *Tridens flava*
Purpletop

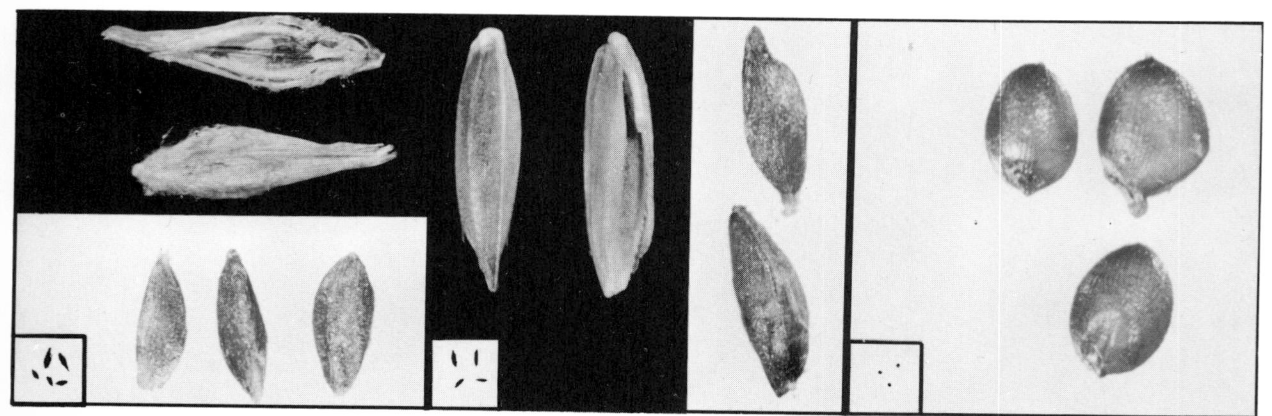

7. *Poa annua*
Annual Bluegrass

8. *Poa pratensis*
Kentucky Bluegrass

9. *Eragrostis cilianensis*
Stinkgrass

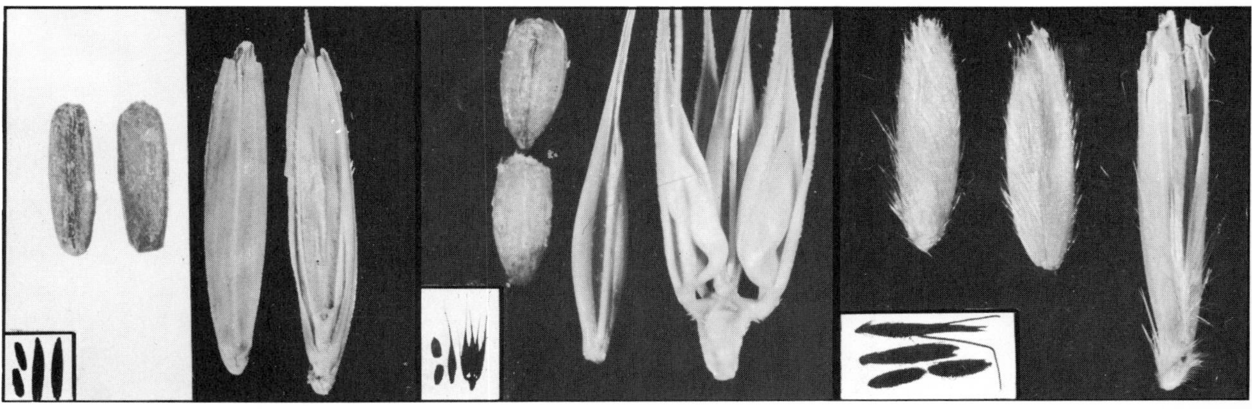

10. *Agropyron repens*
Quackgrass

11. *Hordeum pusillum*
Little Barley

12. *Avena fatua*
Wild Oats

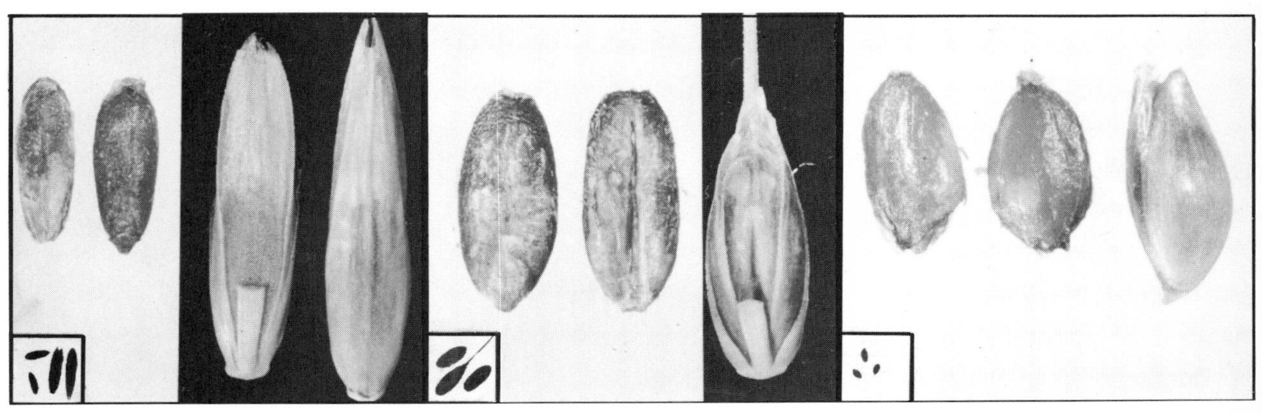

13. *Lolium perenne*
Ryegrass

14. *Lolium temulentum*
Ryegrass

15. *Holcus lanatus*
Velvetgrass

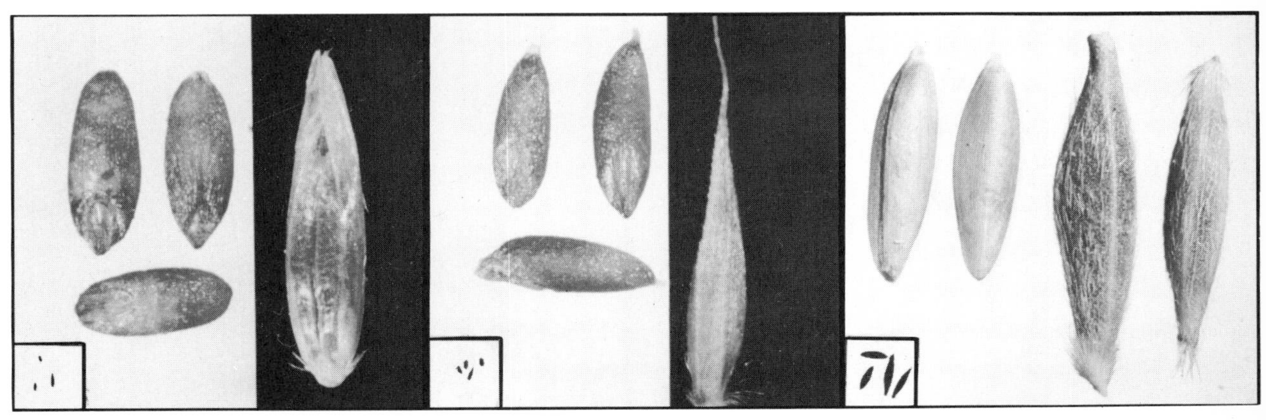

16. *Agrostis alba*
Redtop

17. *Muhlenbergia schreberi*
Nimblewill

18. *Stipa viridula*
Needlegrass

19. *Sporobolus cryptandrus*
Dropseed

20. *Sporobolus asper*
Dropseed

21. *Aristida virgata*
Three-awn

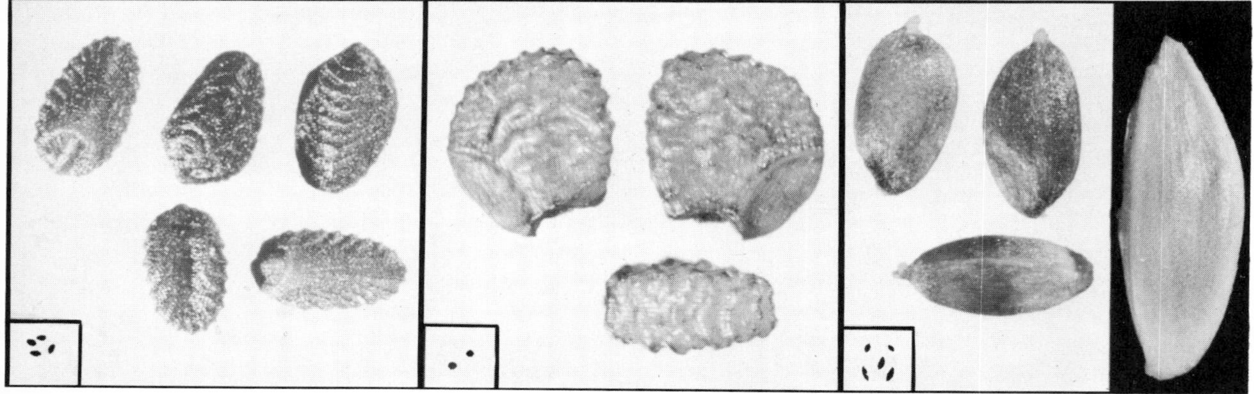

22. *Eleusine indica*
Goosegrass

23. *Dactyloctenium aegyptium*
Crowfootgrass

24. *Cynodon dactylon*
Bermudagrass

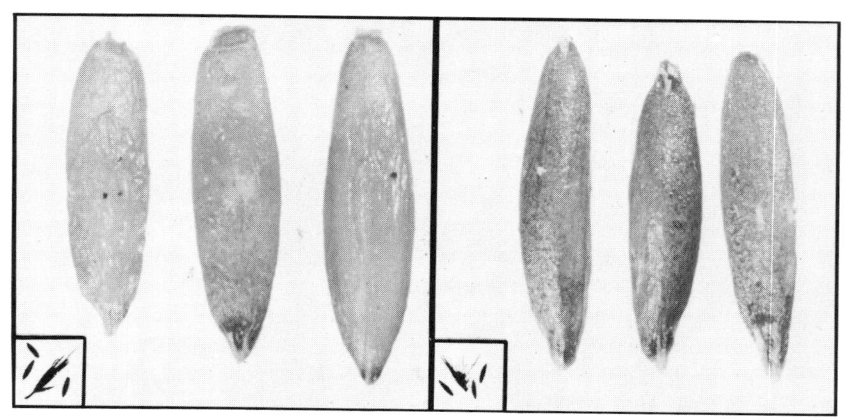

25. *Bouteloua curtipendula*
Grama

26. *Bouteloua gracilis*
Grama

27. *Digitaria filiformis*
Crabgrass

28. *Digitaria ischaemum*
Crabgrass

29. *Digitaria sanguinalis*
Crabgrass

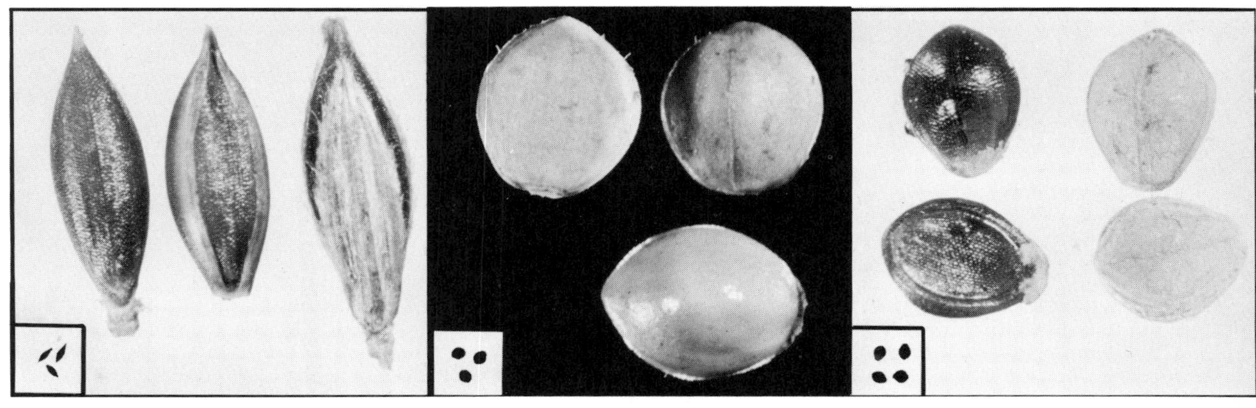

30. *Leptoloma cognatum*
Leptoloma

31. *Paspalum ciliatifolium*
Paspalum

32. *Paspalum boscianum*
Bullgrass

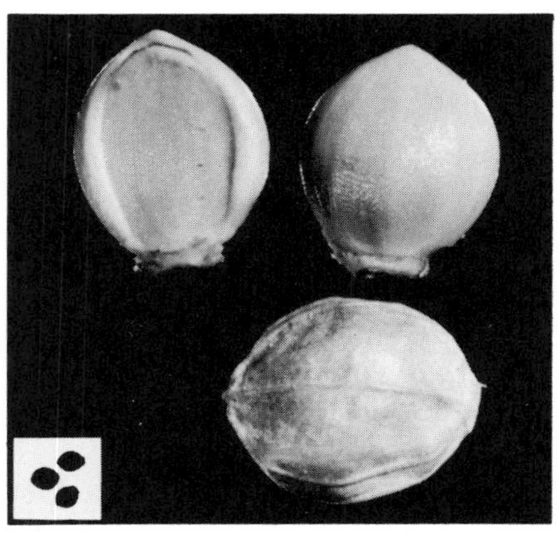

33. *Paspalum laeve*
Paspalum

34. *Panicum capillare*
Witchgrass

35. *Panicum dichotomiflorum*
Panicum

36. *Panicum lindheimeri*
Panicum

37. *Panicum texanum*
Panicum

38. *Setaria viridis*
Bristlegrass

39. *Setaria verticillata*
Bristlegrass

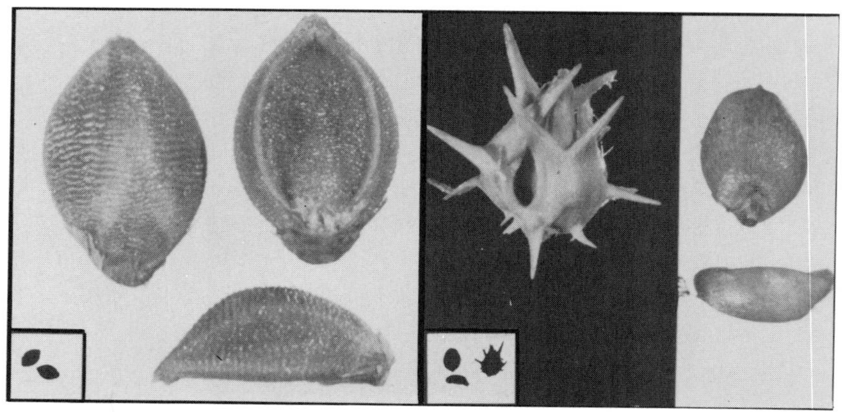

40. *Setaria lutescens*
Bristlegrass

41. *Cenchrus pauciflorus*
Sandbur

42. *Andropogon virginicus*
Broomsedge

43. *Andropogon scoparius*
Bluestem

44. *Sorghum halepense*
Johnsongrass

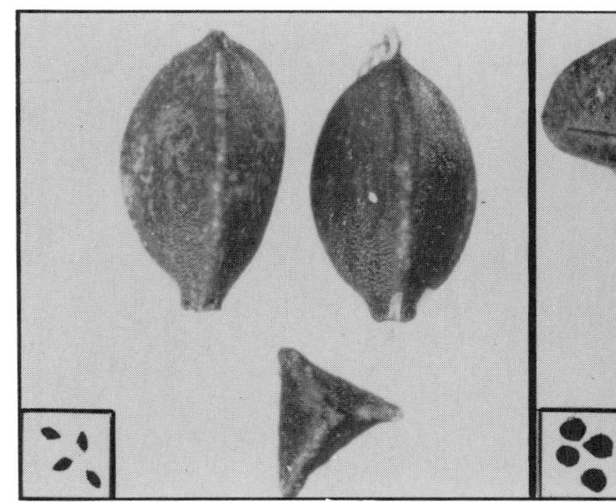

45. *Cyperus schweinitzii*
Flatsedge

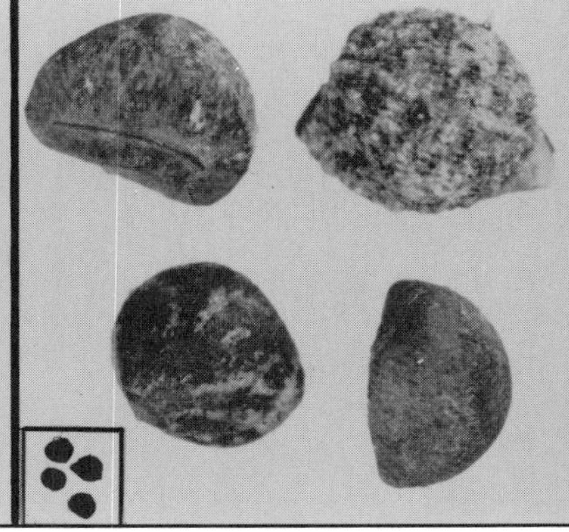

46. *Commelina crispa*
Dayflower

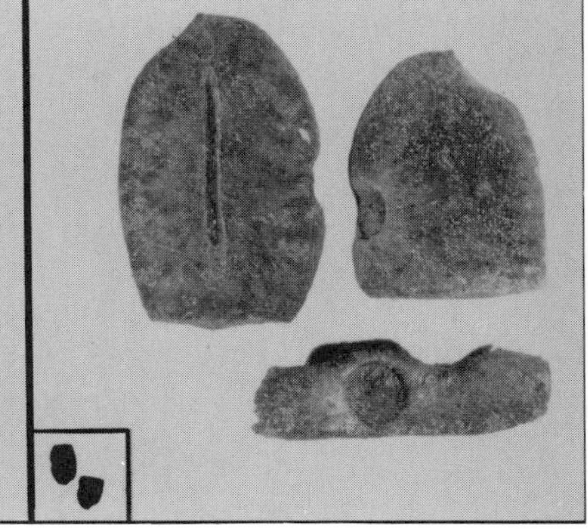

47. *Commelina virginica*
Dayflower

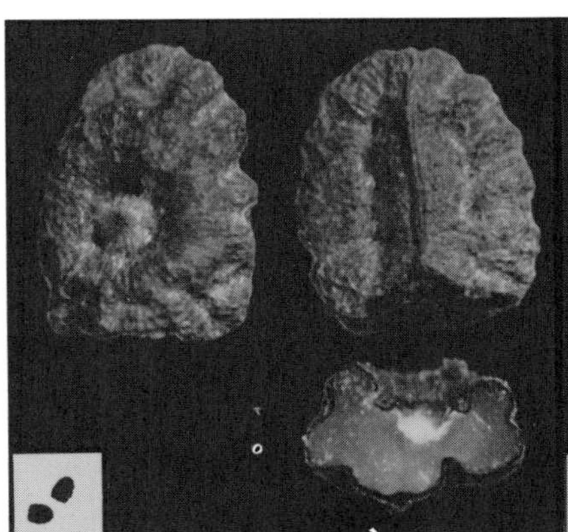

48. *Tradescantia virginiana*
Spiderwort

49. *Eriogonum fasciculatum*
Eriogonum

50. *Eriogonum alatum*
Eriogonum

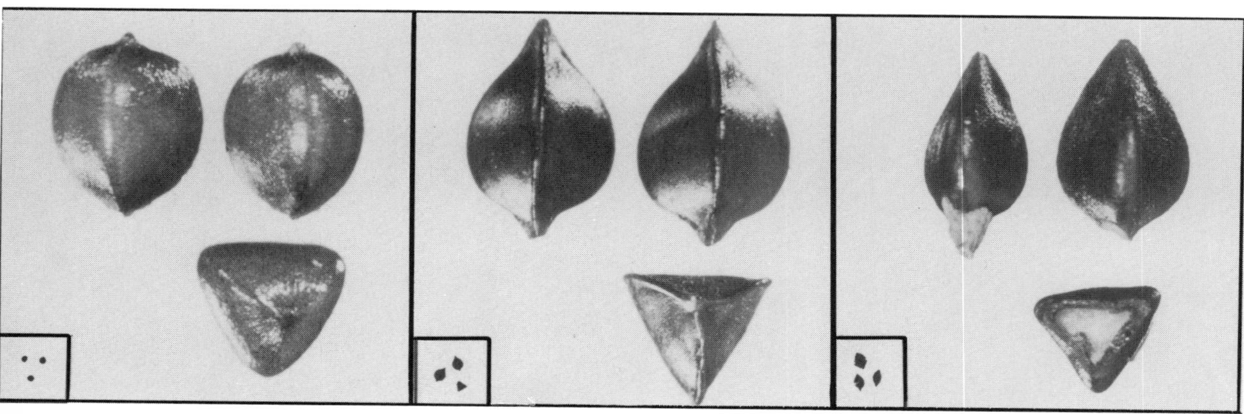

51. *Rumex acetosella*
Sheep Sorrel

52. *Rumex crispus*
Dock

53. *Polygonum aviculare*
Knotweed

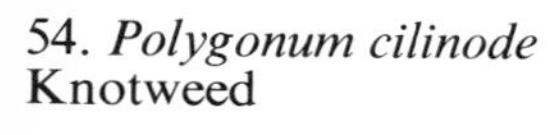

54. *Polygonum cilinode*
Knotweed

55. *Polygonum convolvulus*
Cornbind

Additional species of *Polygonum* are illustrated in plates 631–642 and in figures 71–76

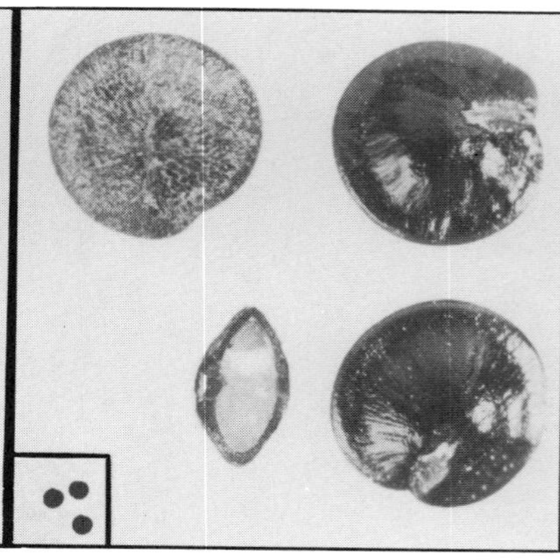

56. *Chenopodium ambrosioides*
Wormseed

57. *Chenopodium album*
Lambsquarters

58. *Chenopodium hybridum*
Goosefoot

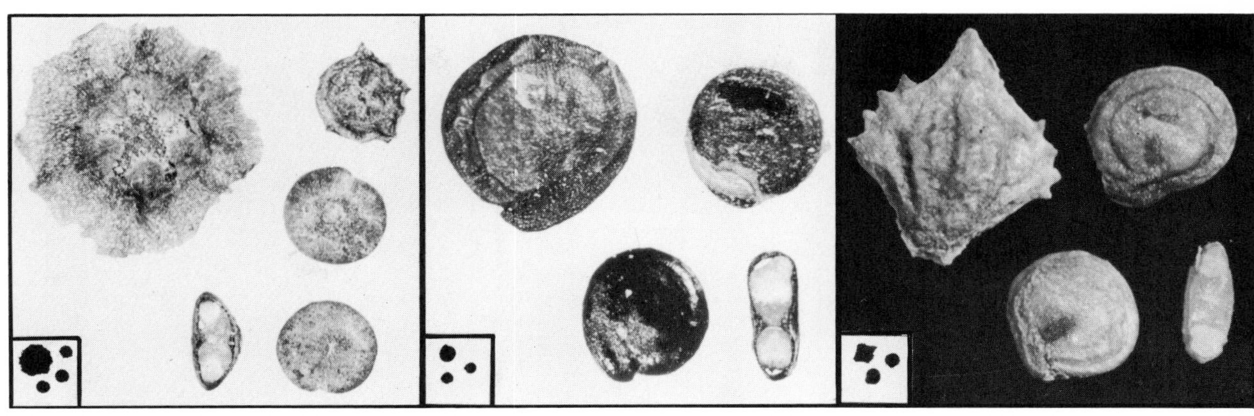

59. *Cycloloma atriplicifolium*
Ringwing

60. *Atriplex patula*
Saltbush

61. *Atriplex semibaccata*
Saltbush

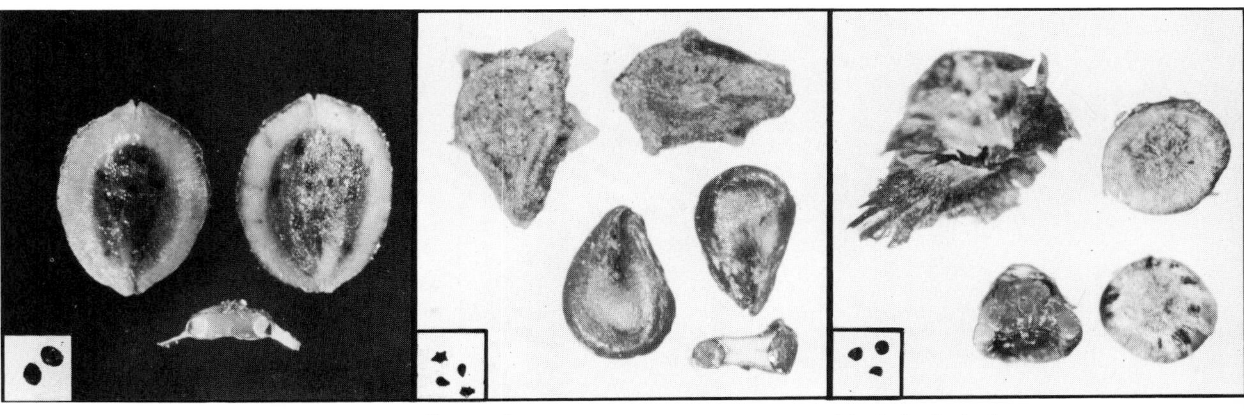

62. *Corispermum hyssopifolium*
Tickseed

63. *Kochia scoparia*
Summer-cypress

64. *Salsola kali*
Russianthistle

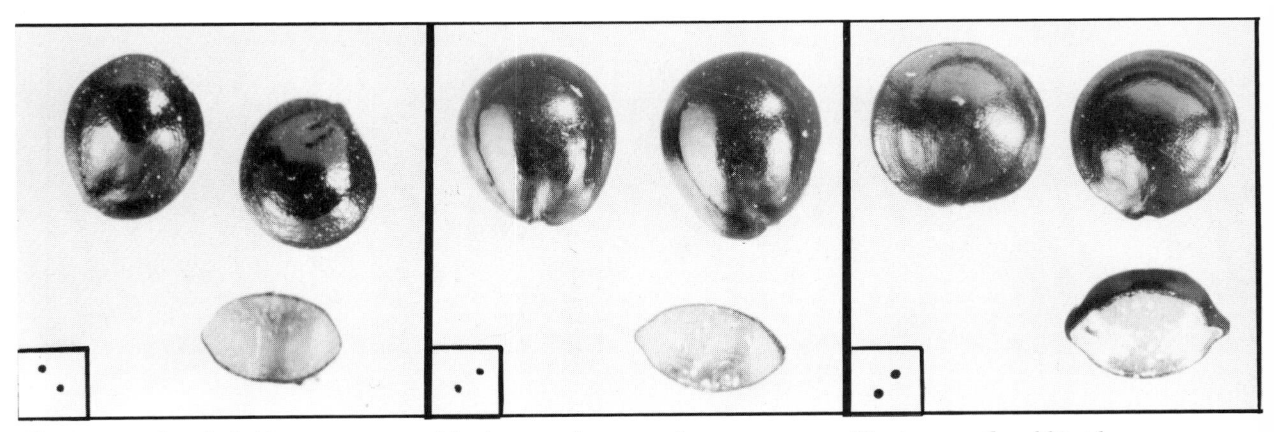

65. *Amaranthus hybridus*
Pigweed

66. *Amaranthus retroflexus*
Pigweed

67. *Amaranthus blitoides*
Pigweed

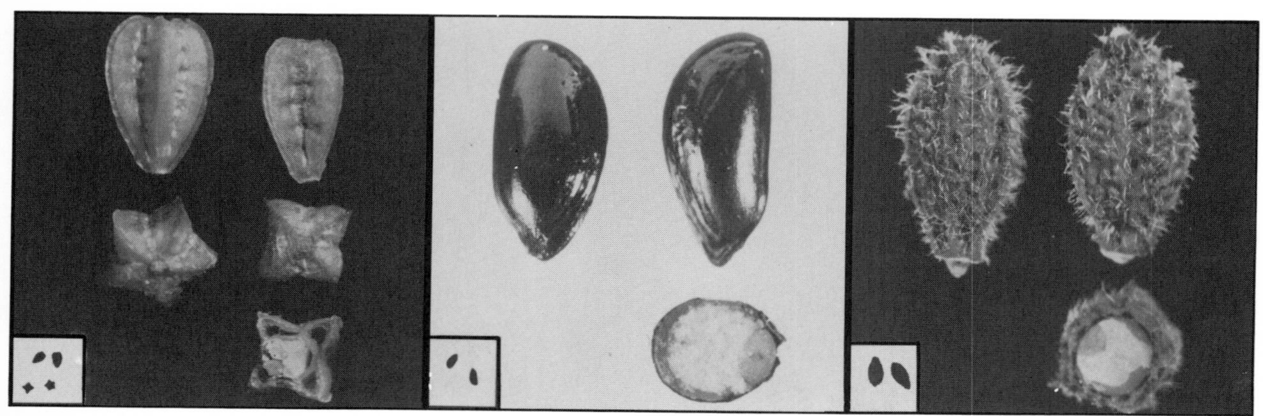

68. *Boerhaavia wrightii*
Spiderling

69. *Abronia fragrans*
Sandverbena

70. *Oxybaphus nyctagineus*
Umbrellawort

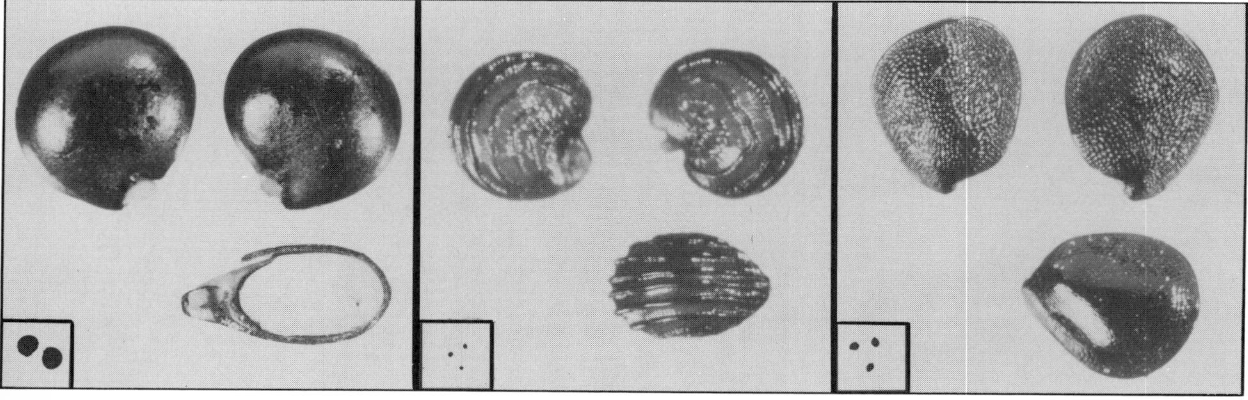

71. *Phytolacca americana*
Pokeberry

72. *Mollugo verticillata*
Carpetweed

73. *Calandrinia caulescens*
Redmaids

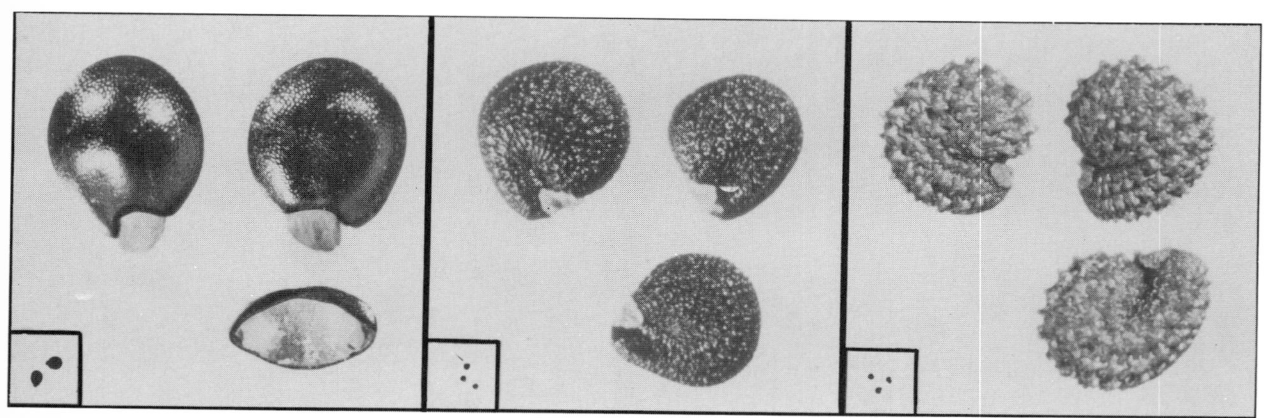

74. *Montia perfoliata*
Minerslettuce

75. *Portulaca oleracea*
Purslane

76. *Portulaca lanceolata*
Purslane

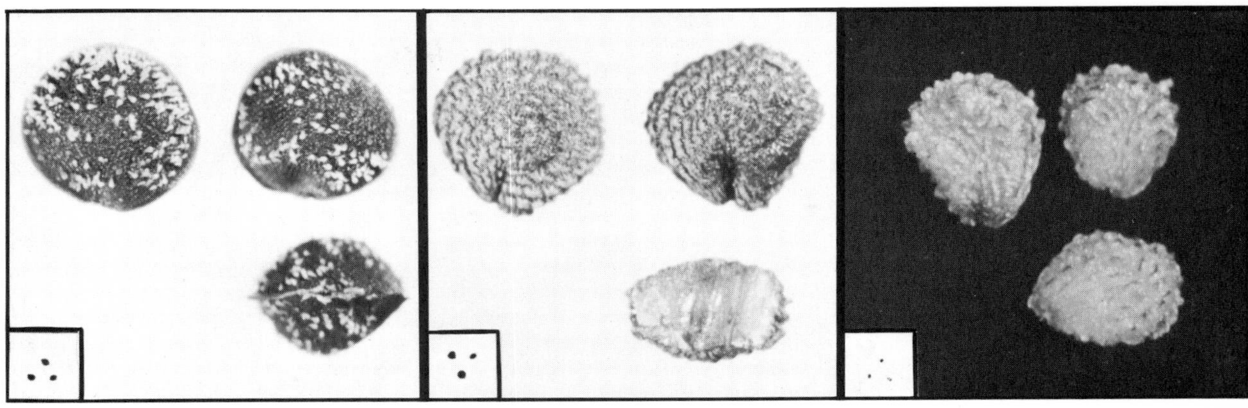

77. *Spergula arvensis*
Spurry

78. *Stellaria media*
Chickweed

79. *Cerastium viscosum*
Chickweed

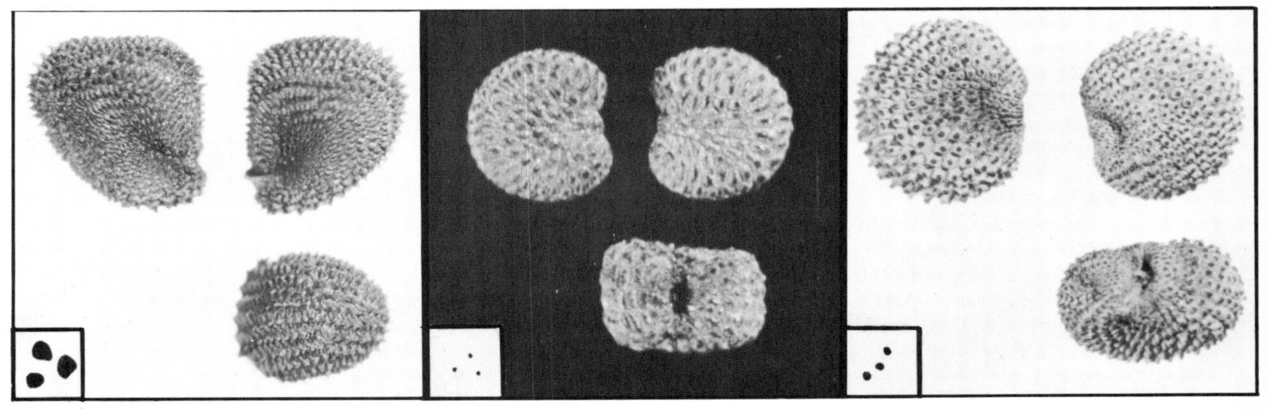

80. *Agrostemma githago*
Cockle

81. *Silene antirrhina*
Catchfly

82. *Silene noctiflora*
Catchfly

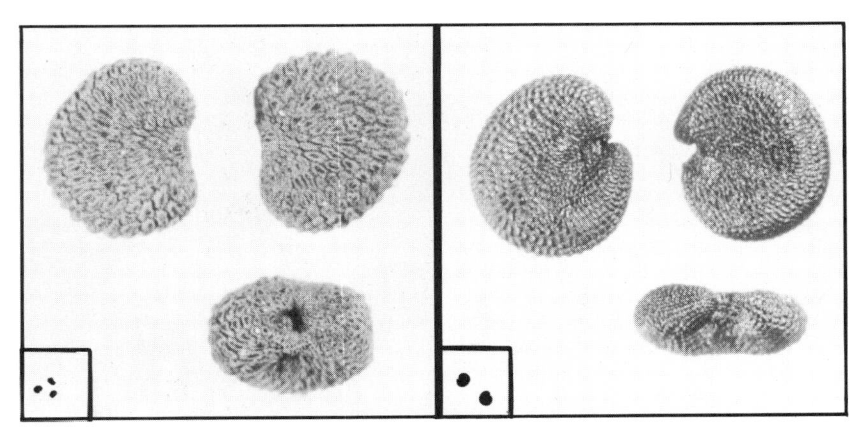

83. *Lychnis coronaria*
Campion

84. *Saponaria officinalis*
Bouncingbet

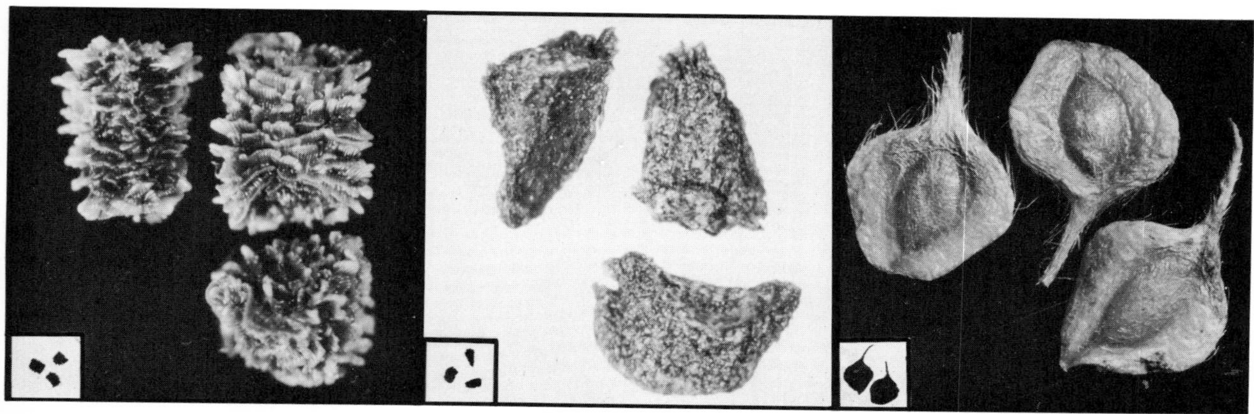

85. *Delphinium virescens*
Larkspur

86. *Delphinium cardinale*
Larkspur

87. *Anemone canadensis*
Anemone

88. *Berberis vulgaris*
Barberry

89. *Berberis aquifolium*
Oregongrape

90. *Argemone platyceras*
Pricklypoppy

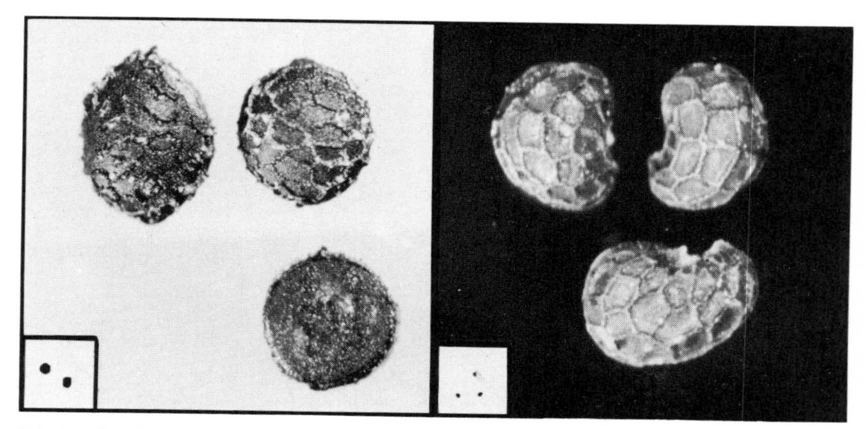

91. *Eschscholtzia californica*
California-poppy

92. *Papaver dubium*
Poppy

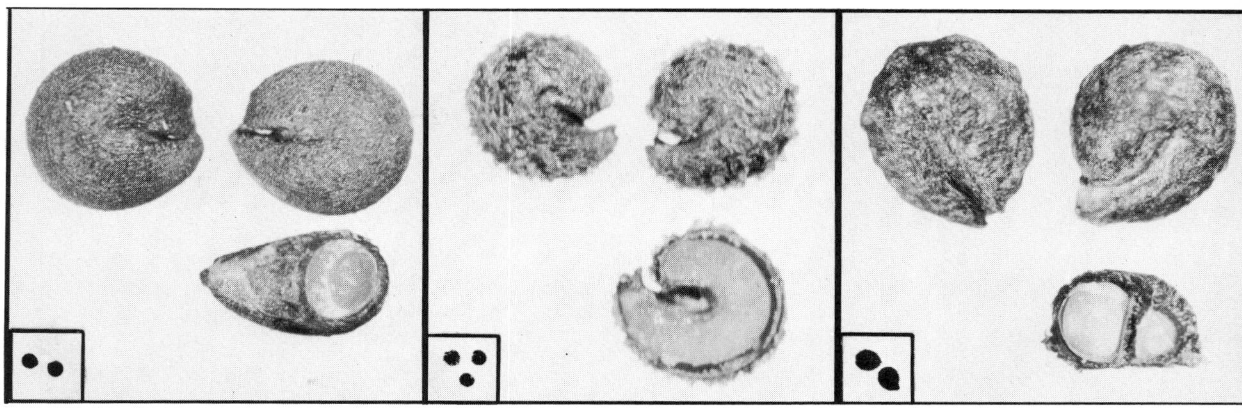

93. *Polanisia graveolens*
Clammyweed

94. *Cleome spinosa*
Spiderflower

95. *Cleome serrulata*
Spiderflower

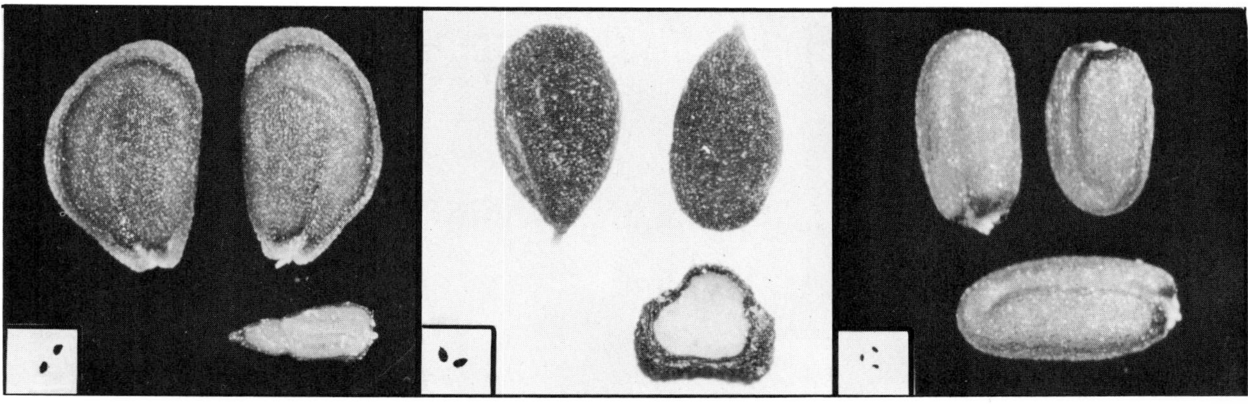

96. *Lepidium virginicum*
Pepperweed

97. *Lepidium campestre*
Pepperweed

98. *Capsella bursa-pastoris*
Shepherds-purse

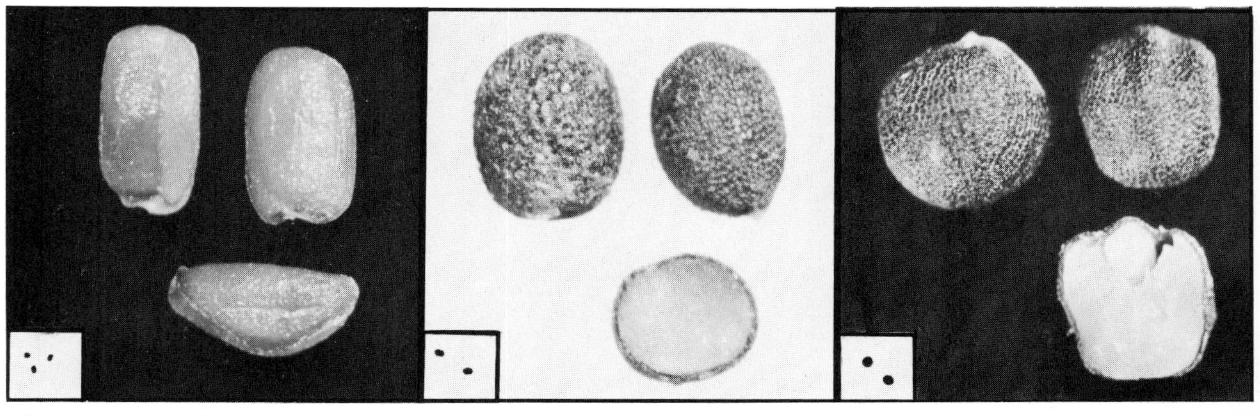

99. *Sisymbrium altissimum*
Tumblemustard

100. *Brassica nigra*
Mustard

101. *Brassica campestris*
Bird Rape

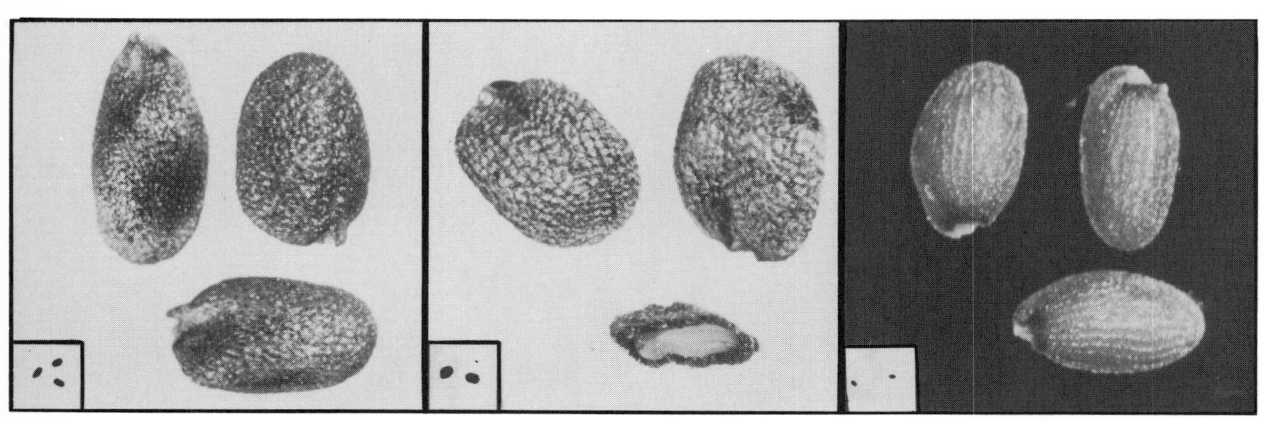

102. *Barbarea vulgaris*
Wintercress

103. *Barbarea verna*
Wintercress

104. *Descurainia pinnata*
Tansymustard

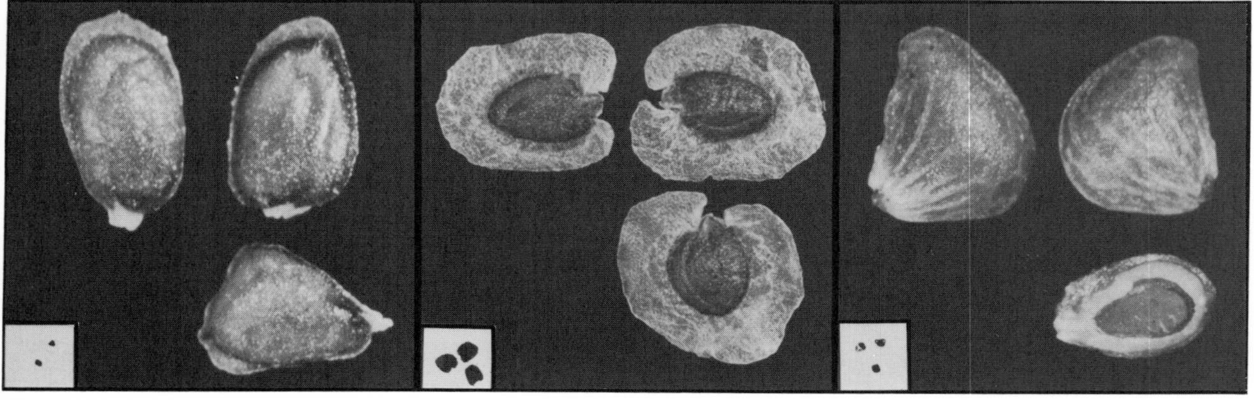

105. *Arabis glabra*
Rockcress

106. *Arabis canadensis*
Rockcress

107. *Fragaria virginiana*
Strawberry

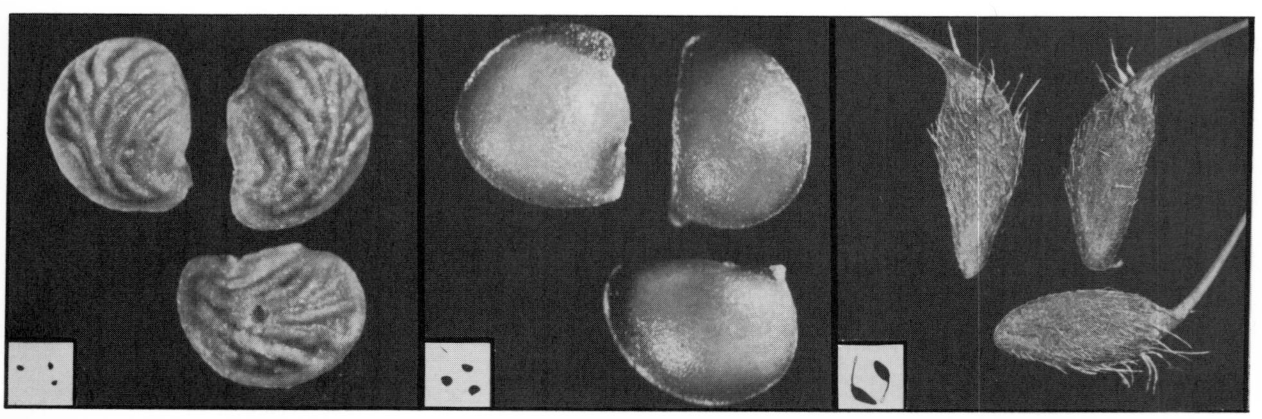

108. *Potentilla pensylvanica*
Cinquefoil

109. *Potentilla palustris*
Cinquefoil

110. *Geum canadense*
Avens

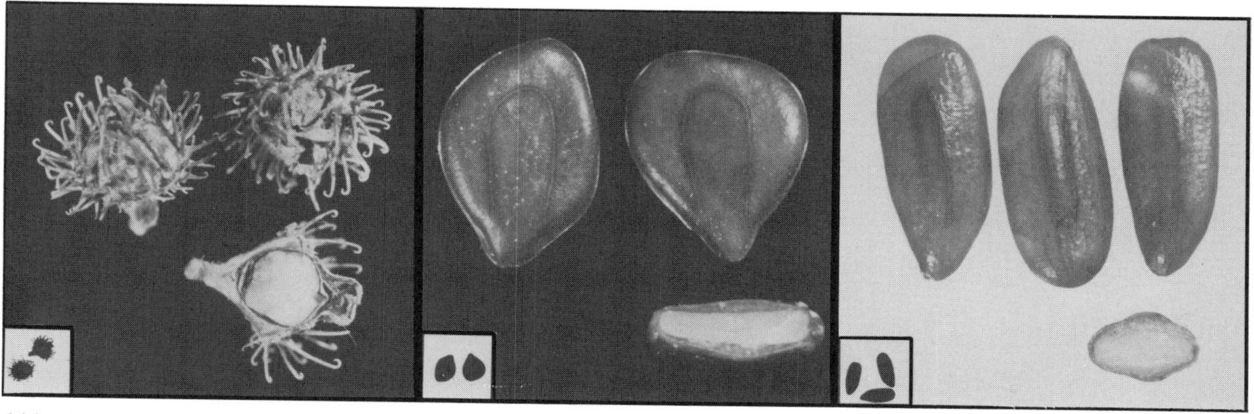

111. *Agrimonia parviflora*
Agrimony

112. *Desmanthus illinoensis*
Bundleflower

113. *Desmanthus leptolobus*
Bundleflower

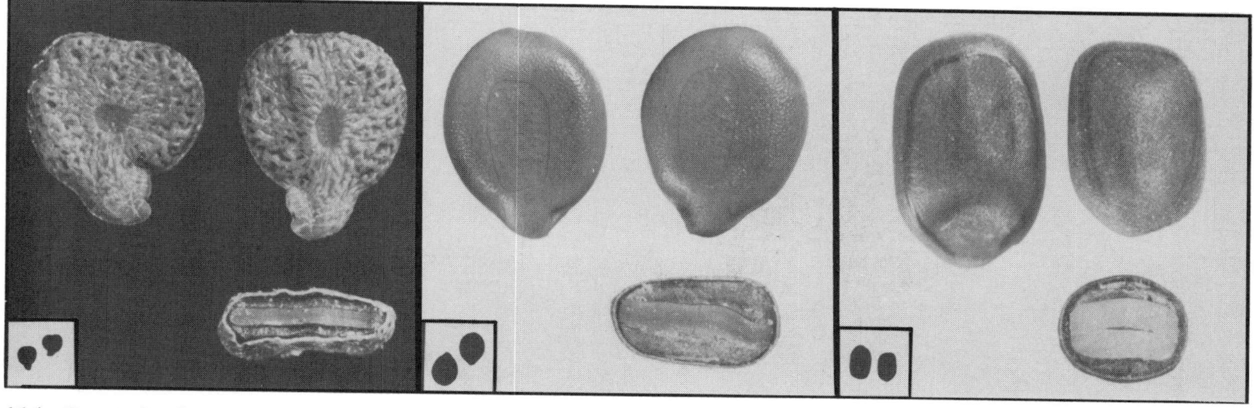

114. *Cassia bauhinioides*
Senna

115. *Cassia occidentalis*
Senna

116. *Schrankia angustata*
Sensitivebrier

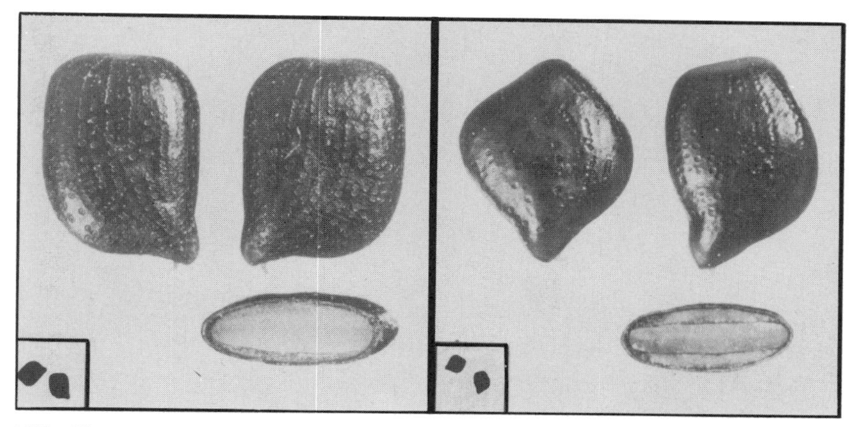

117. *Chamaecrista fasciculata*
Partridgepea

118. *Chamaecrista nictitans*
Partridgepea

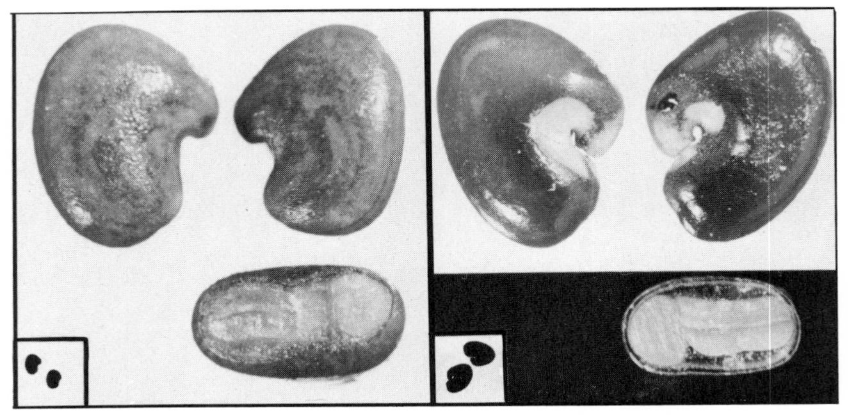

119. *Crotalaria striata*
Crotalaria

120. *Crotalaria spectabilis*
Crotalaria

121. *Lupinus sparsiflorus*
Lupine

122. *Lupinus texensis*
Lupine

123. *Lupinus perennis*
Lupine

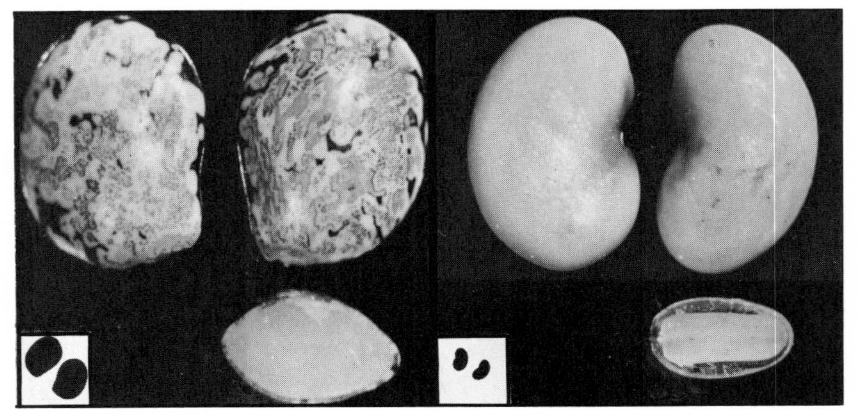

124. *Lupinus albifrons*
Lupine

125. *Medicago hispida*
Burclover

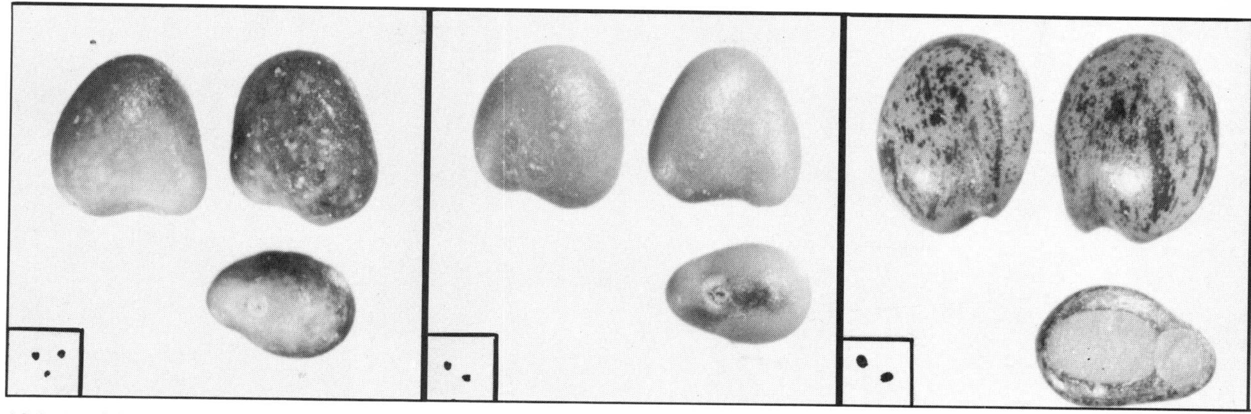

126. *Trifolium repens*
White Clover

127. *Trifolium hybridum*
Alsike

128. *Trifolium tridentatum*
Clover

129. *Trifolium ciliatum*
Clover

130. *Trifolium obtusiflorum*
Clover

131. *Lotus tomentellus*
Deervetch

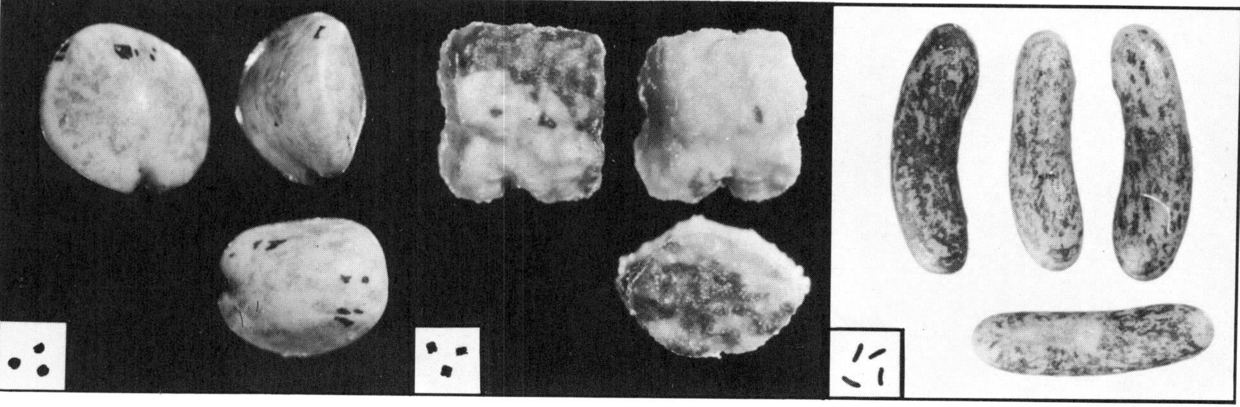

132. *Lotus humistratus*
Deervetch

133. *Lotus strigosus*
Deervetch

134. *Lotus scoparius*
Deervetch

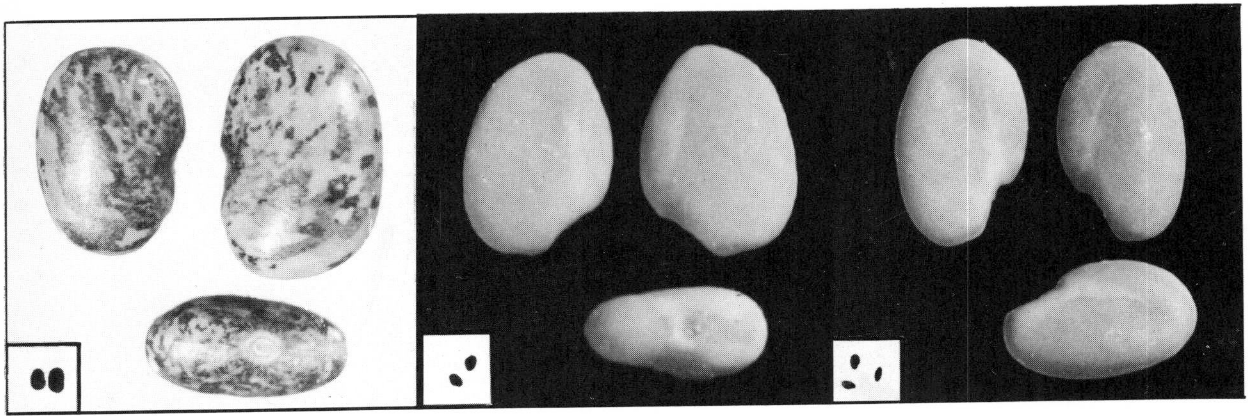

135. *Lotus americanus*
Deervetch

136. *Melilotus alba*
Sweetclover

137. *Melilotus officinalis*
Sweetclover

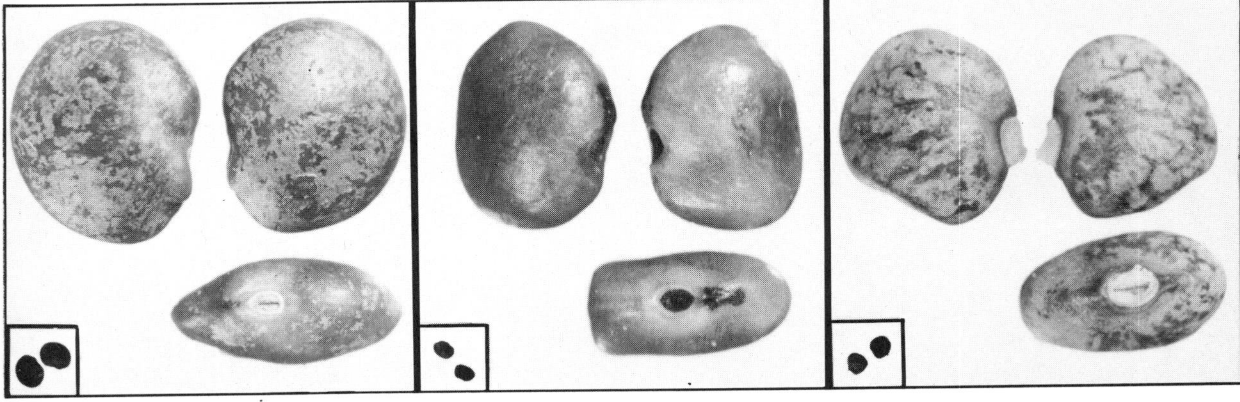

138. *Psoralea canescens*
Scurfpea

139. *Indigofera caroliniana*
Indigo

140. *Tephrosia spicata*
Hoarypea

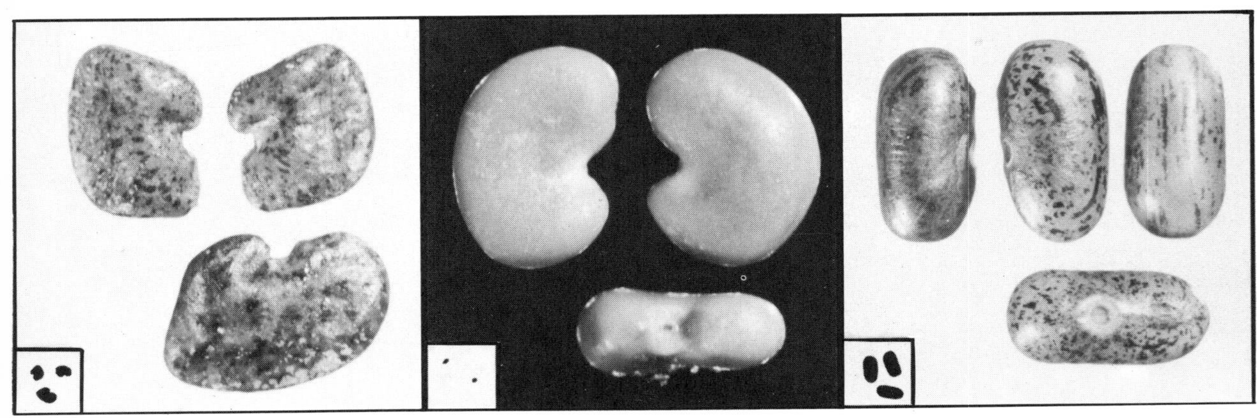

141. *Astragalus nuttallianus*
Loco

142. *Astragalus canadensis*
Loco

143. *Sesbania macrocarpa*
Coffeeweed

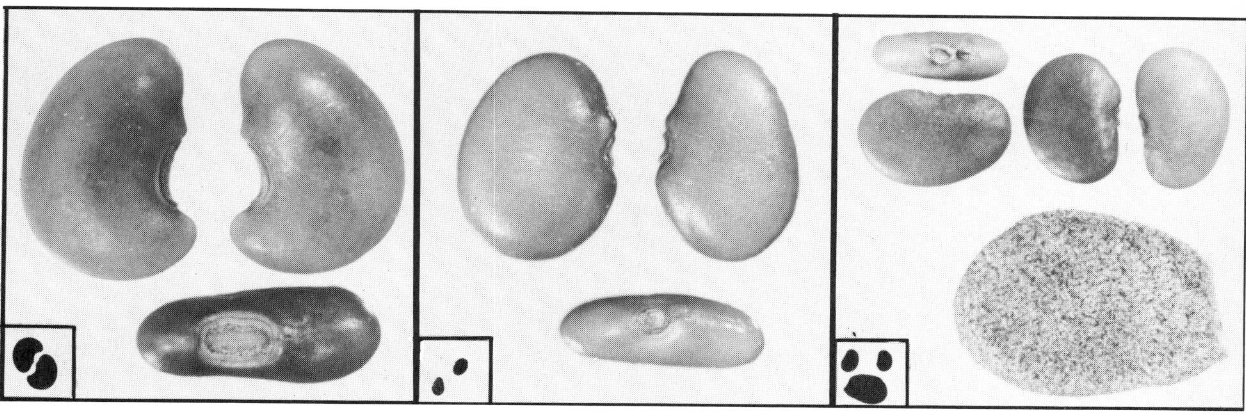

144. *Aeschynomene virginica*
Jointvetch

145. *Desmodium purpureum*
Tickclover

146. *Desmodium canadense*
Tickclover

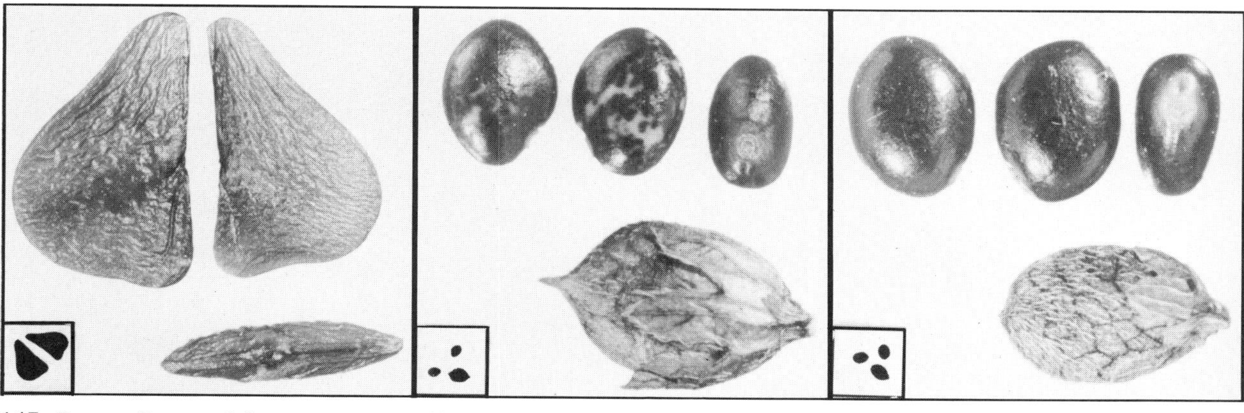

147. *Desmodium nudiflorum*
Tickclover

148. *Lespedeza striata*
Lespedeza

149. *Lespedeza stipulacea*
Lespedeza

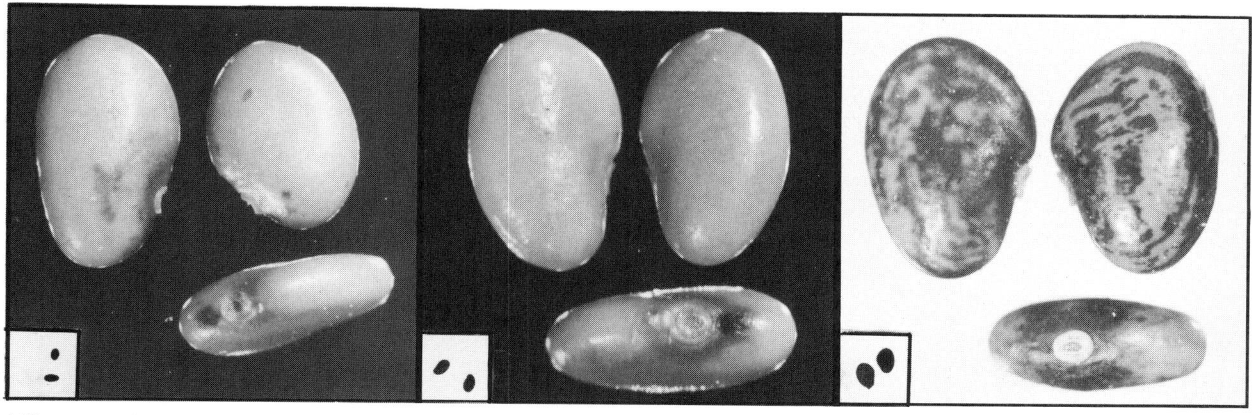

150. *Lespedeza cuneata*
Lespedeza

151. *Lespedeza capitata*
Lespedeza

152. *Lespedeza bicolor*
Lespedeza

153. *Vicia angustifolia*
Vetch

154. *Vicia sylvatica*
Vetch

155. *Lathyrus pusillus*
Peavine

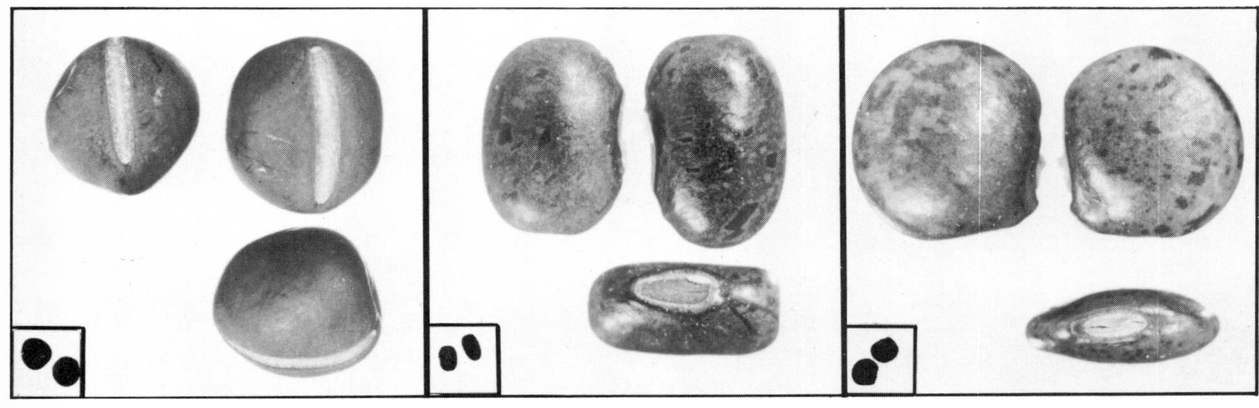

156. *Lathyrus maritimus*
Peavine

157. *Centrosema virginianum*
Butterflypea

158. *Rhynchosia erecta*
Rhynchosia

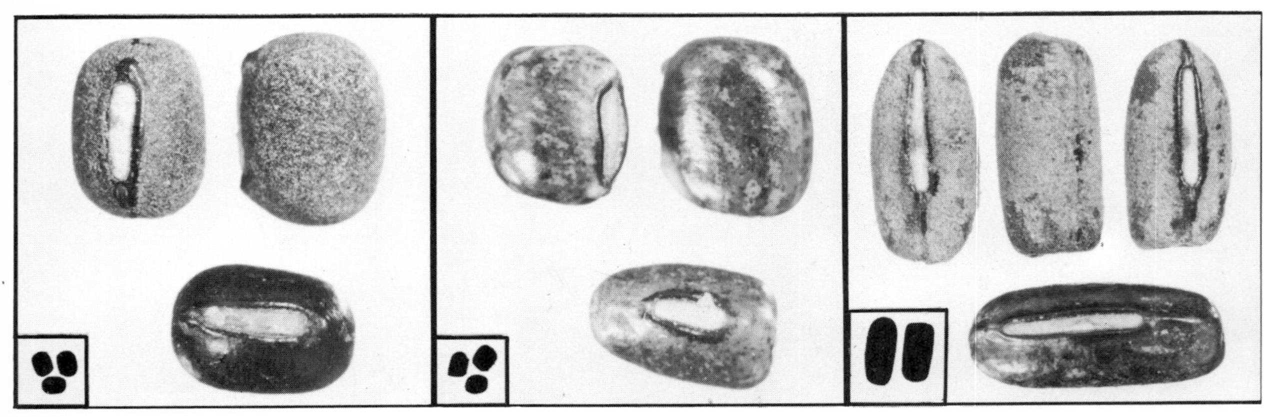

159. *Strophostyles umbellata*
Wildbean

160. *Strophostyles leiosperma*
Wildbean

161. *Strophostyles helvola*
Wildbean

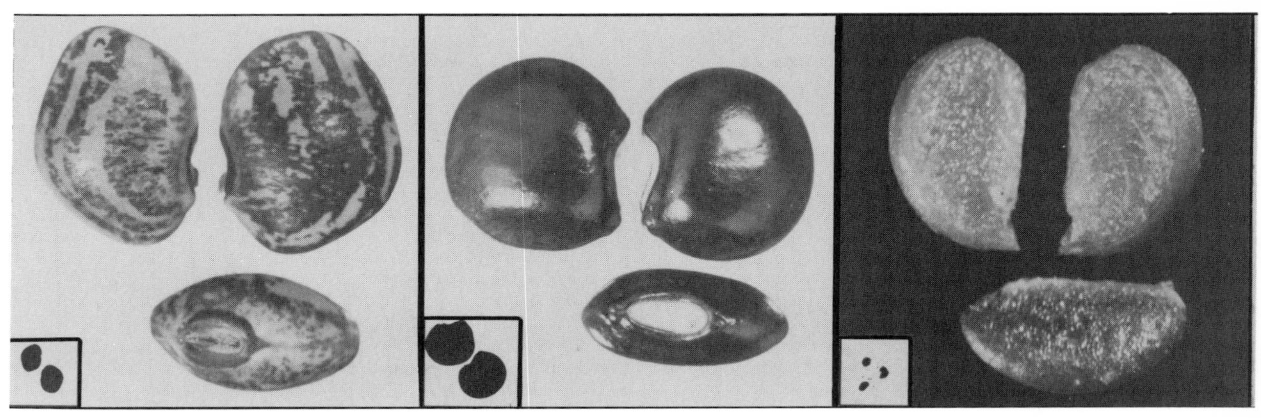

162. *Galactia regularis*
Milkpea

163. *Phaseolus polystachyus*
Bean

164. *Linum striatum*
Flax

165. *Oxalis stricta*
Oxalis

166. *Geranium molle*
Geranium

167. *Geranium carolinianum*
Geranium

168. *Erodium cicutarium*
Filaree

169. *Erodium botrys*
Filaree

170. *Tribulus terrestris*
Puncturevine

171. *Kallstroemia grandiflora*
Caltrop

172. *Polygala cruciata*
Milkwort

173. *Polygala longa*
Milkwort

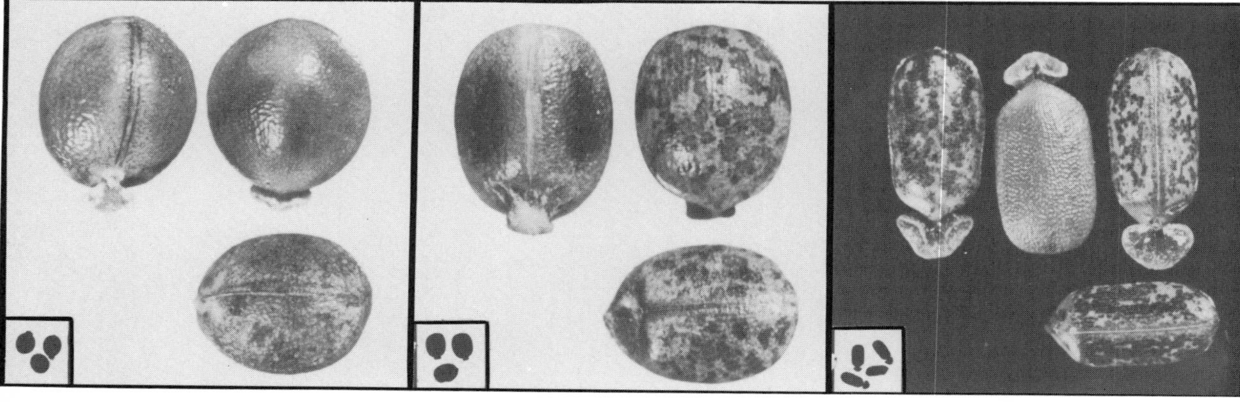

174. *Croton monanthogynus*
Doveweed

175. *Croton glandulosus*
Doveweed

176. *Croton lindheimerianus*
Doveweed

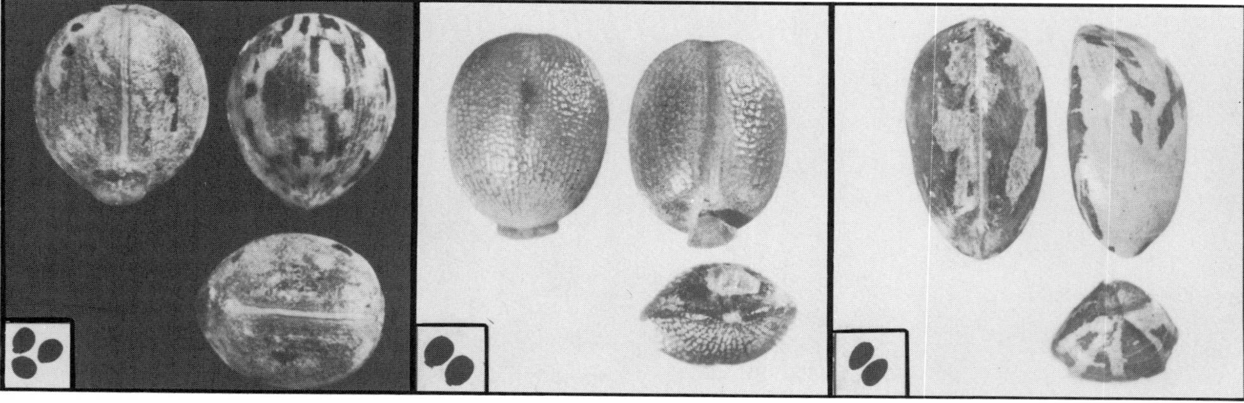

177. *Croton texensis*
Doveweed

178. *Croton capitatus*
Doveweed

179. *Eremocarpus setigerus*
Turkeymullein

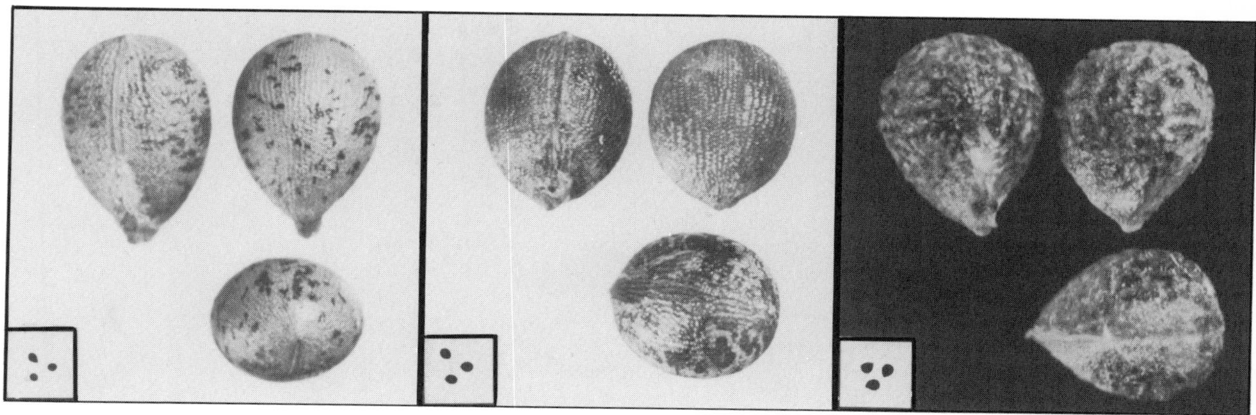

180. *Acalypha virginica*
Copperleaf

181. *Acalypha gracilens*
Copperleaf

182. *Acalypha ostryaefolia*
Copperleaf

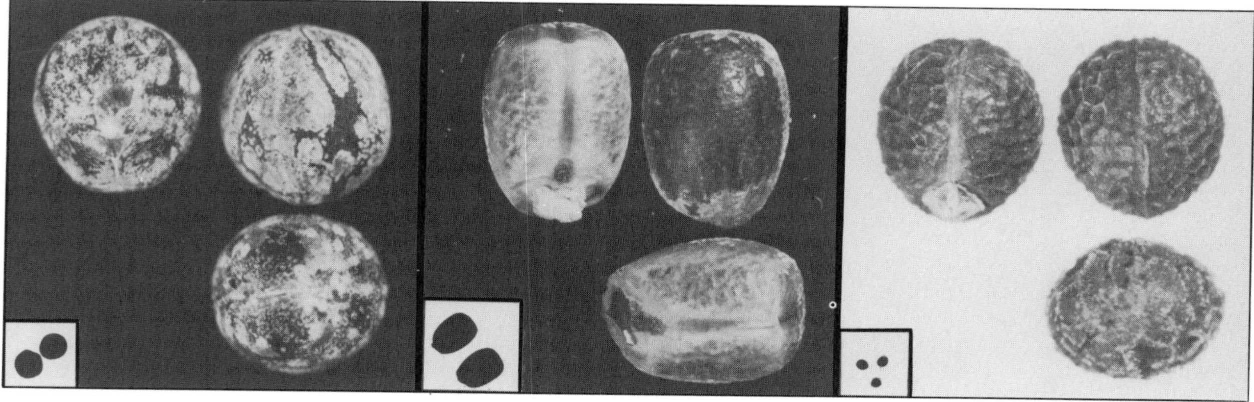

183. *Tragia urens*
Noseburn

184. *Stillingia sylvatica*
Queensdelight

185. *Euphorbia dictyosperma*
Spurge

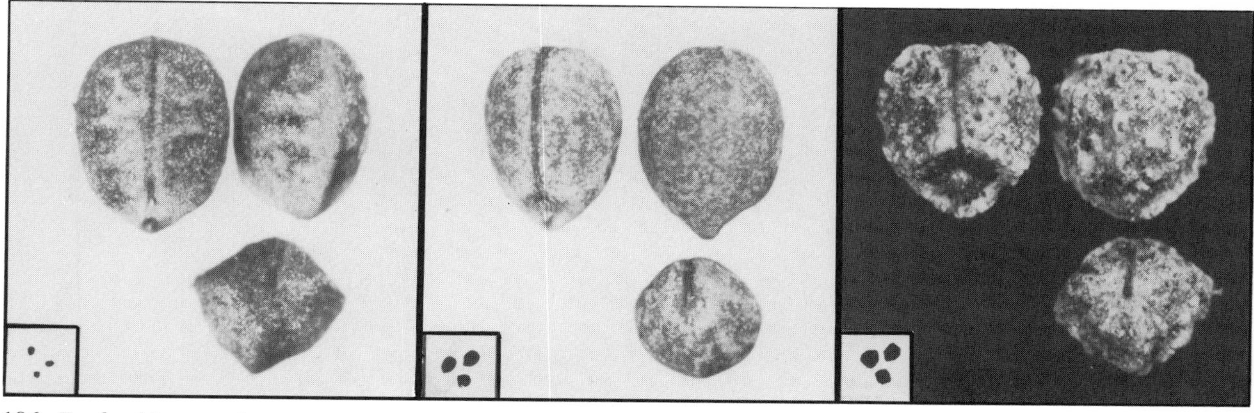

186. *Euphorbia maculata*
Spurge

187. *Euphorbia corollata*
Spurge

188. *Euphorbia dentata*
Spurge

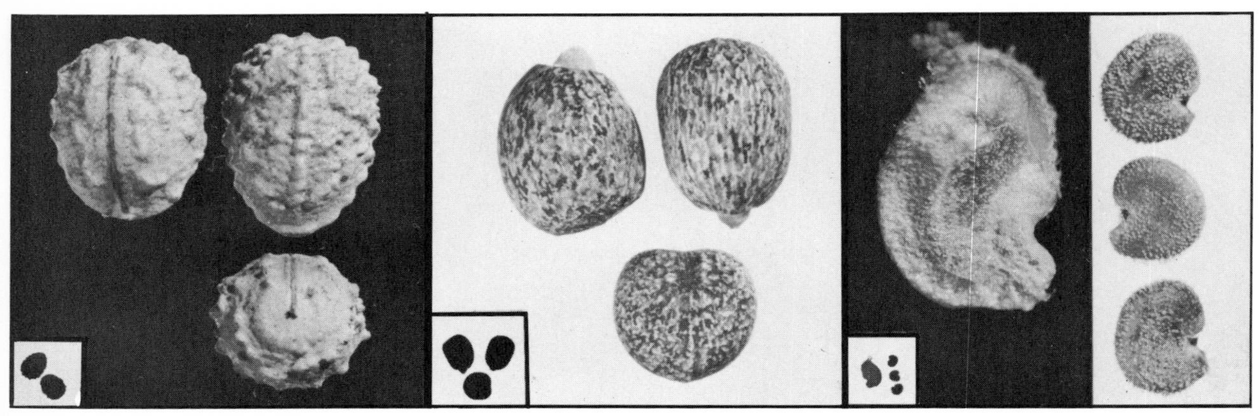

189. *Euphorbia marginata*
Spurge

190. *Euphorbia lathyrus*
Moleweed

191. *Sphaeralcea fendleri*
Globemallow

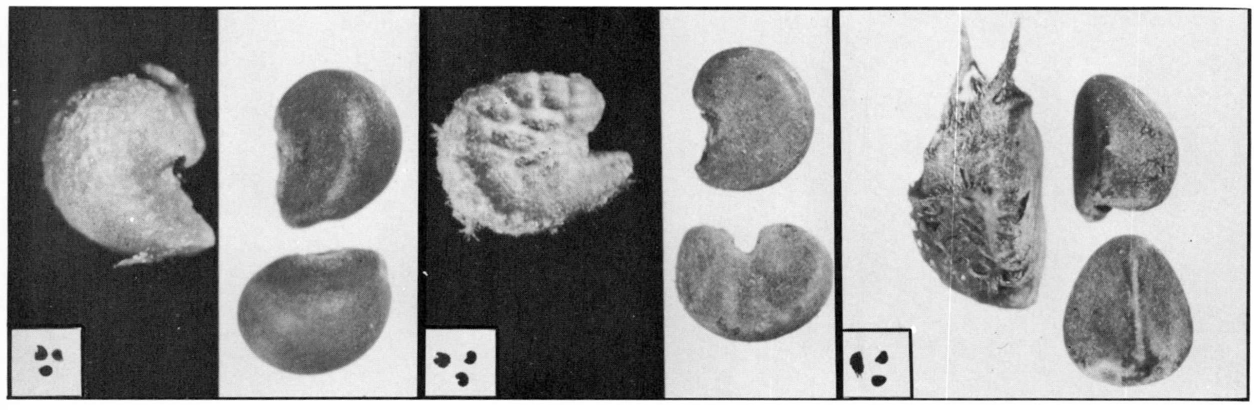

192. *Sidalcea neomexicana*
Checkermallow

193. *Malvastrum coccineum*
Falsemallow

194. *Sida spinosa*
Sida

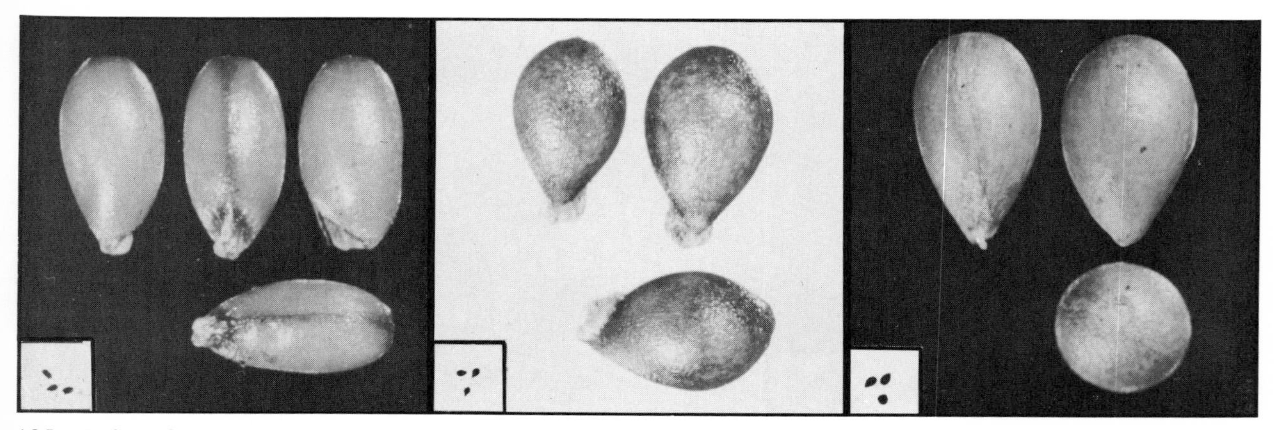

195. *Viola rafinesquii*
Johnny-jump-up

196. *Viola lanceolata*
Violet

197. *Viola palmata*
Violet

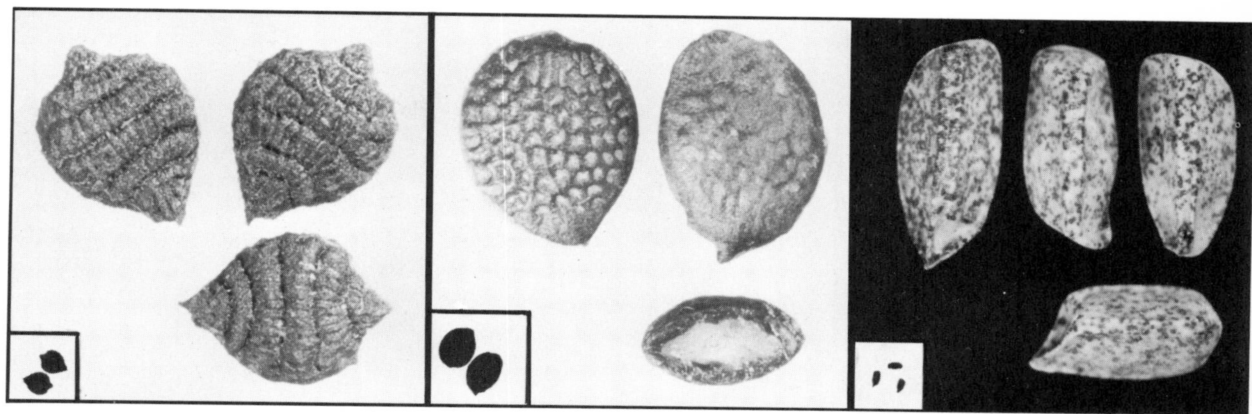

198. *Passiflora lutea*
Passionflower

199. *Passiflora incarnata*
Passionflower

200. *Mentzelia micrantha*
Prairiestar

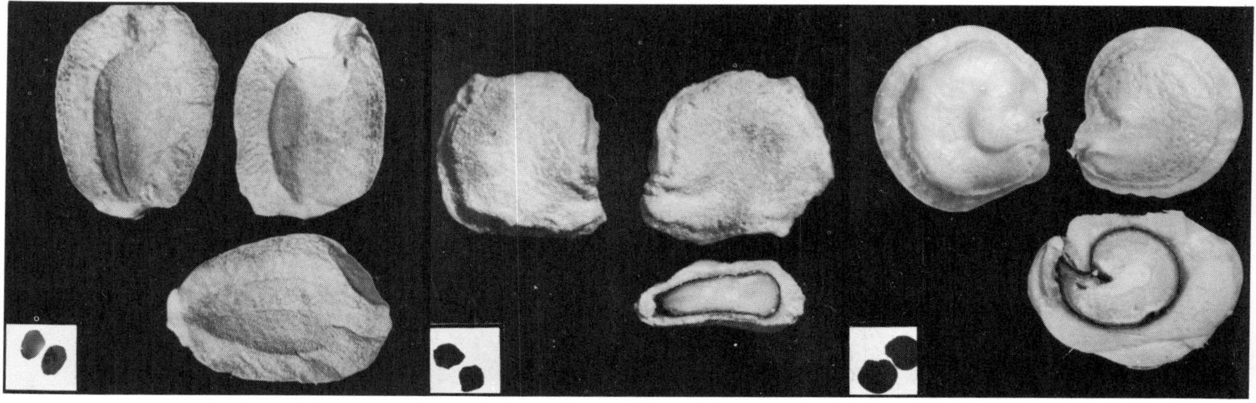

201. *Mentzelia decapetala*
Prairiestar

202. *Opuntia leptocaulis*
Pricklypear

203. *Opuntia polycantha*
Pricklypear

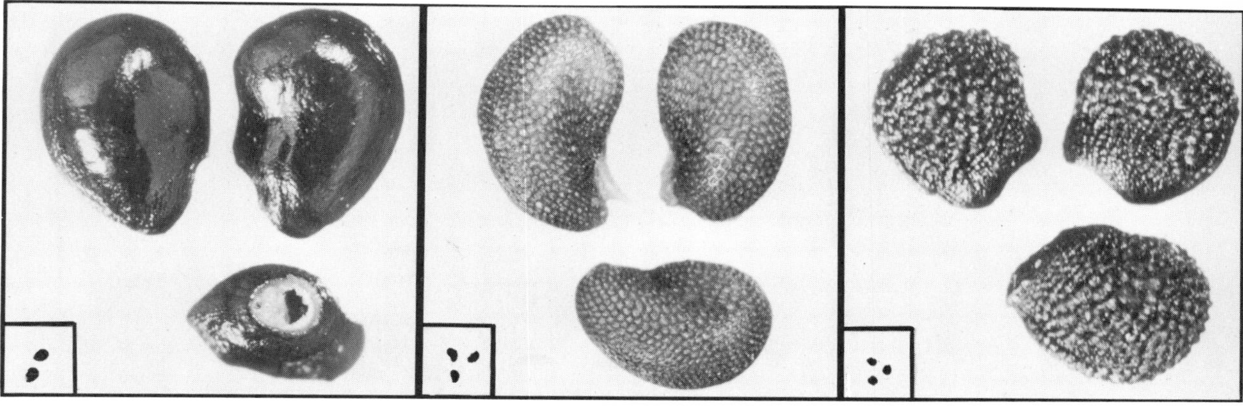

204. *Cereus giganteus*
Giantcactus

205. *Mammillaria vivipara*
Cactus

206. *Echinocactus rigidissimus*
Cactus

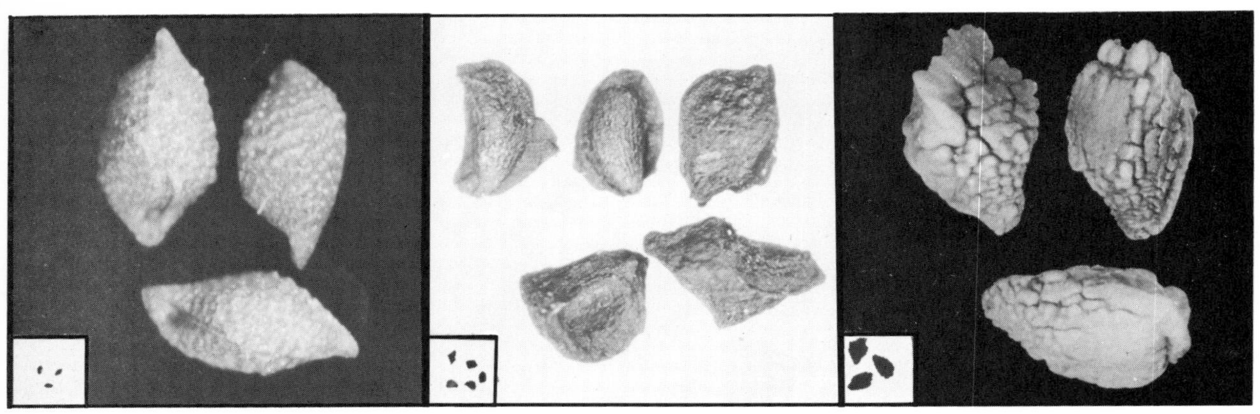

207. *Oenothera laciniata*
Evening-primrose

208. *Oenothera biennis*
Evening-primrose

209. *Oenothera missouriensis*
Sundrops

210. *Gaura biennis*
Gaura

211. *Daucus carota*
Wildcarrot

212. *Asclepias syriaca*
Milkweed

213. *Breweria pickeringii*
Breweria

214. *Convolvulus arvensis*
Bindweed

215. *Convolvulus sepium*
Morning-glory

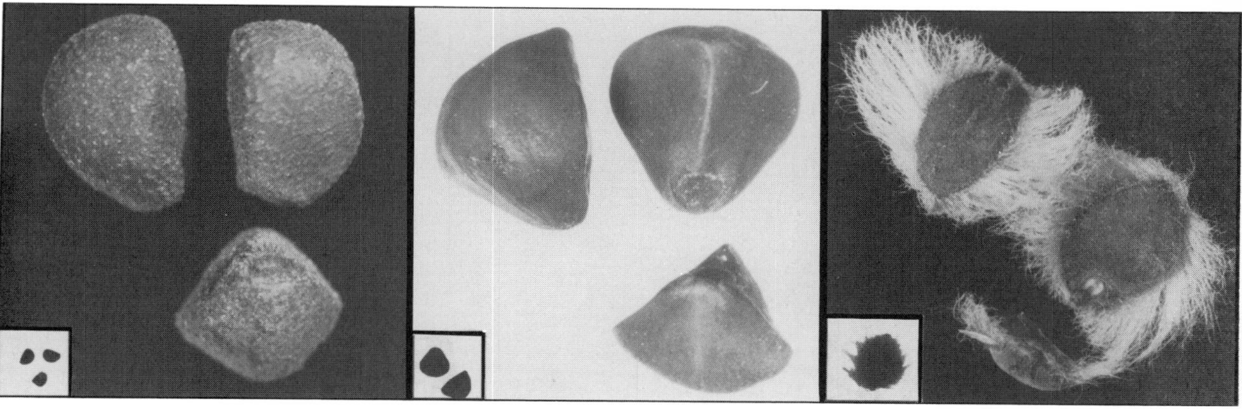

216. *Jacquemontia tamnifolia*
Jacquemontia

217. *Ipomoea lacunosa*
Morning-glory

218. *Ipomoea pandurata*
Morning-glory

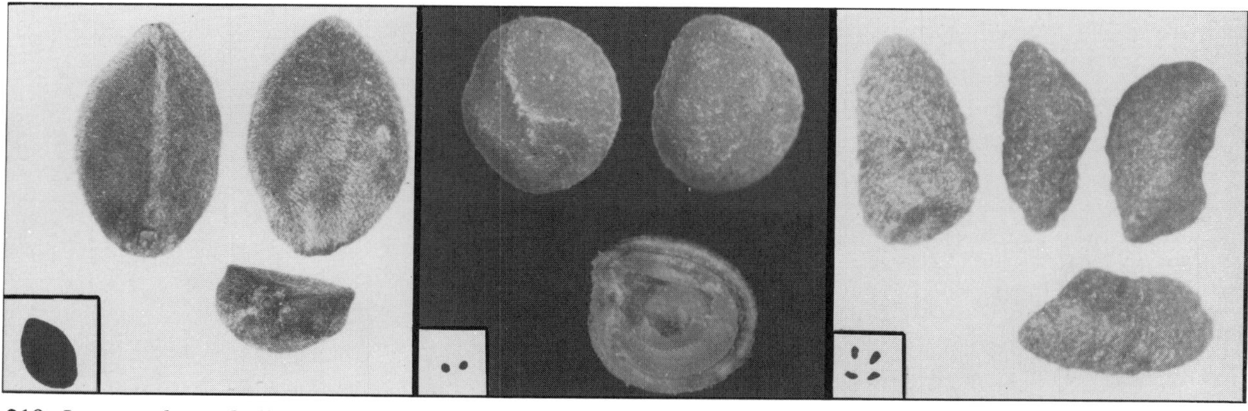

219. *Ipomoea leptophylla*
Morning-glory

220. *Cuscuta gronovii*
Dodder

221. *Gilia tricolor*
Gilia

222. *Gilia aggregata*
Gilia

223. *Collomia grandiflora*
Collomia

224. *Phlox pilosa*
Phlox

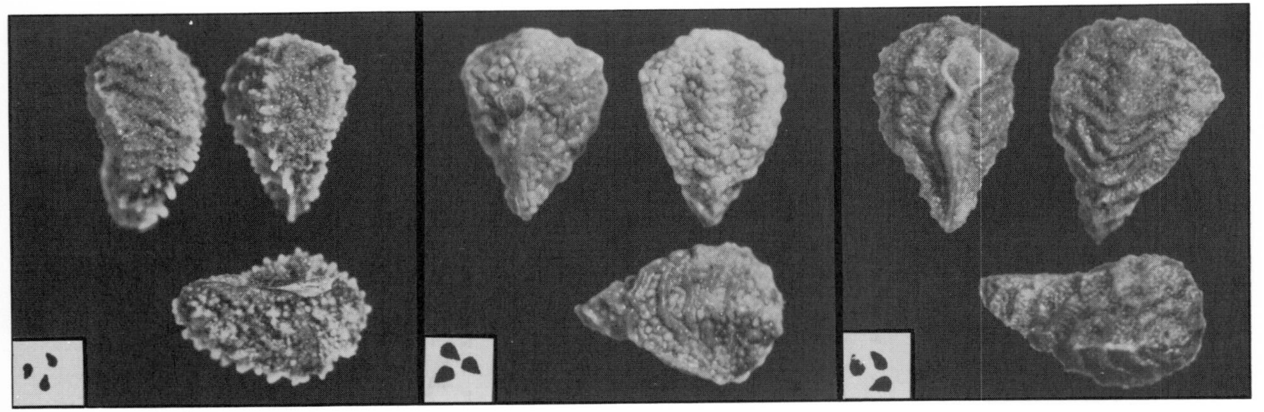

225. *Amsinckia intermedia*
Fiddleneck

226. *Amsinckia tessellata*
Fiddleneck

227. *Amsinckia douglasiana*
Fiddleneck

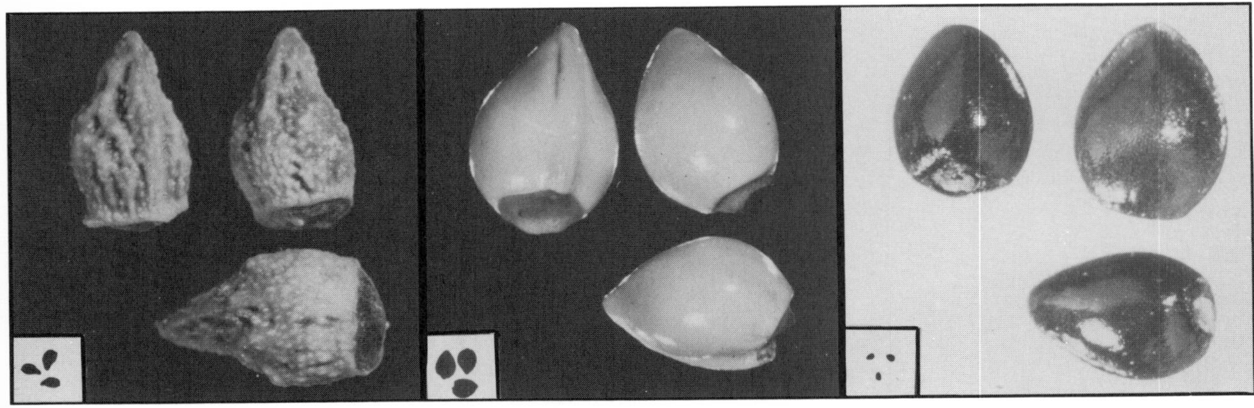

228. *Lithospermum arvense*
Gromwell

229. *Lithospermum carolinense*
Gromwell

230. *Myosotis arvensis*
Forget-me-not

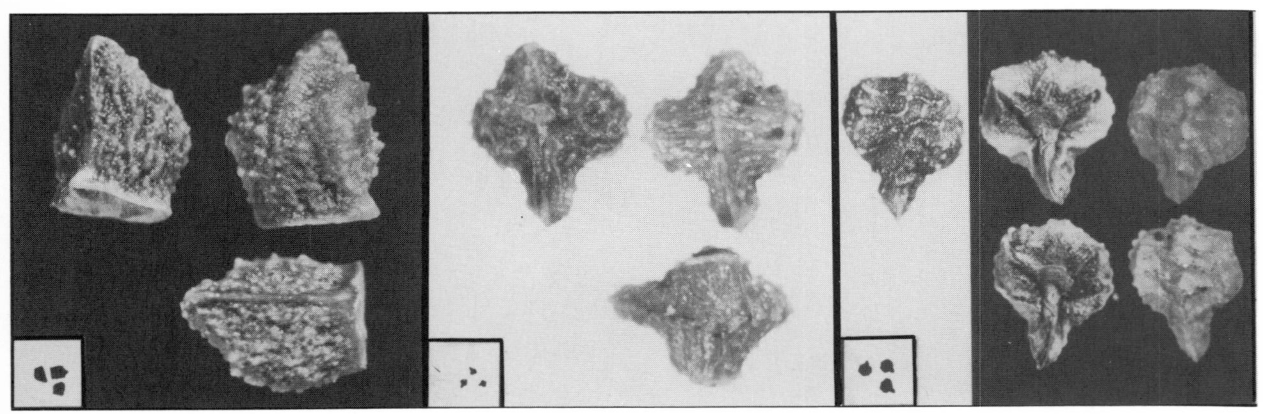

231. *Echium vulgare*
Blueweed

232. *Plagiobothrys tenellus*
Popcornflower

233. *Plagiobothrys nothofulvus*
Popcornflower

234. *Verbena bracteosa*
Verbena

235. *Verbena hastata*
Verbena

236. *Verbena stricta*
Verbena

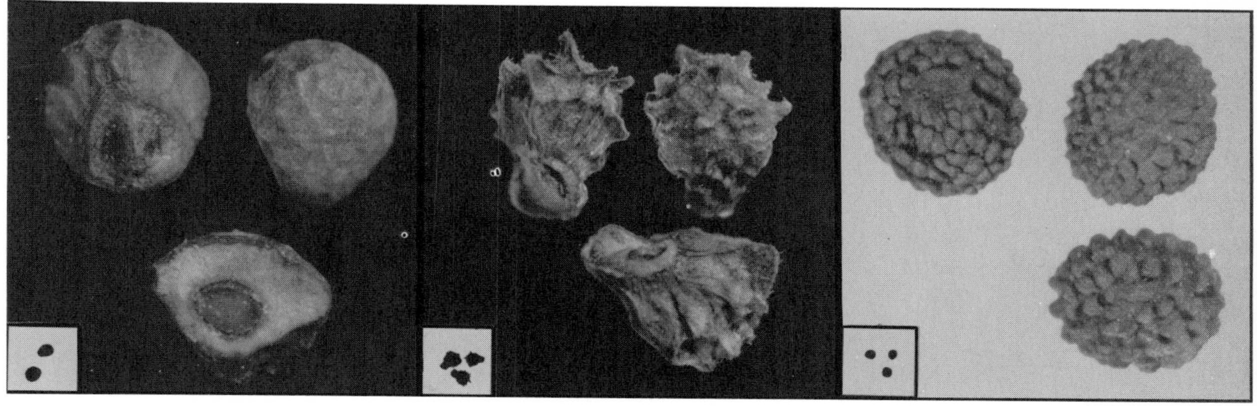

237. *Teucrium canadense*
Germander

238. *Trichostema lanceolatum*
Bluecurls

239. *Scutellaria integrifolia*
Skullcap

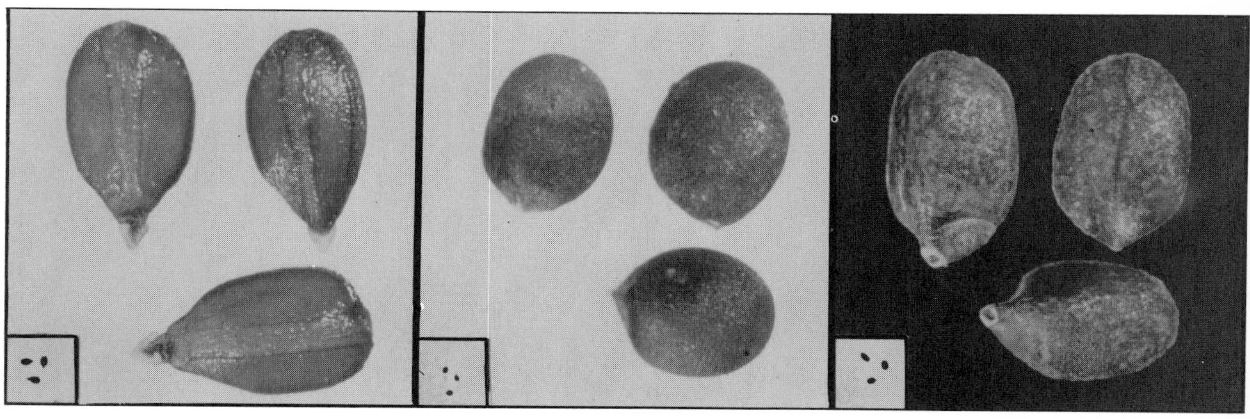

240. *Prunella vulgaris*
Selfheal

241. *Hedeoma pulegioides*
False Pennyroyal

242. *Monarda punctata*
Beebalm

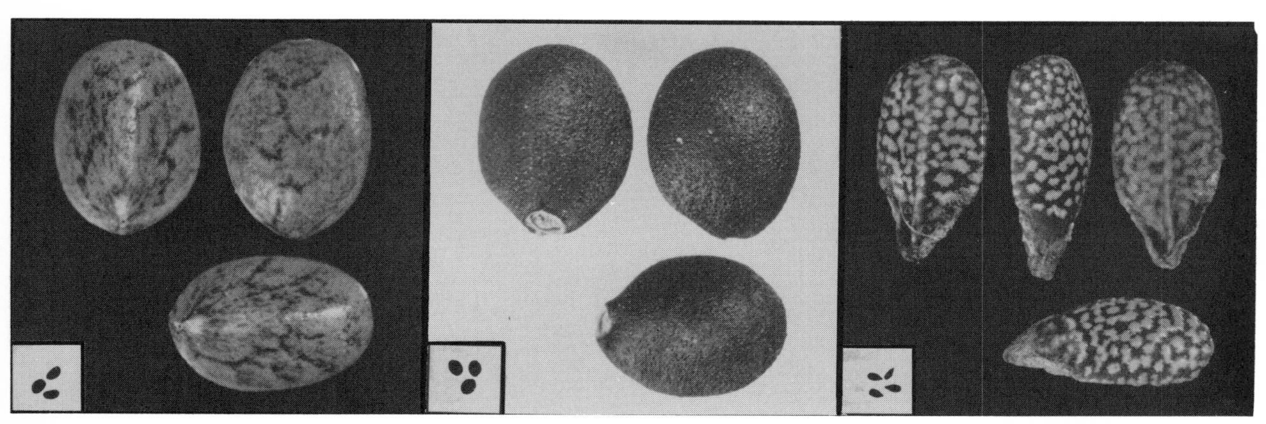

243. *Salvia lanceafolia*
Sage

244. *Salvia lyrata*
Sage

245. *Lamium amplexicaule*
Henbit

246. *Solanum nigrum*
Nightshade

247. *Solanum carolinense*
Horsenettle

248. *Solanum dulcamara*
Nightshade

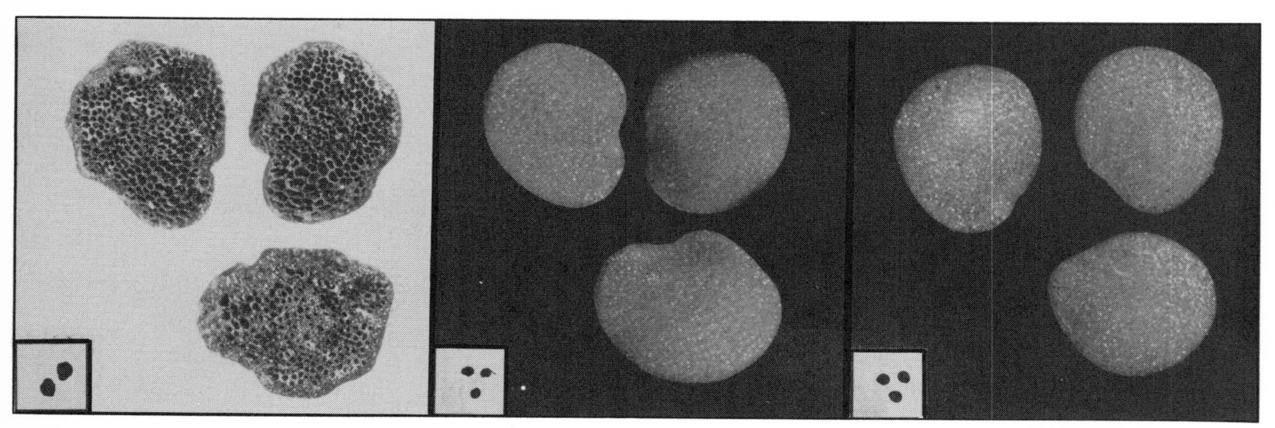

249. *Solanum rostratum*
Buffalobur

250. *Physalis virginiana*
Groundcherry

251. *Physalis heterophylla*
Groundcherry

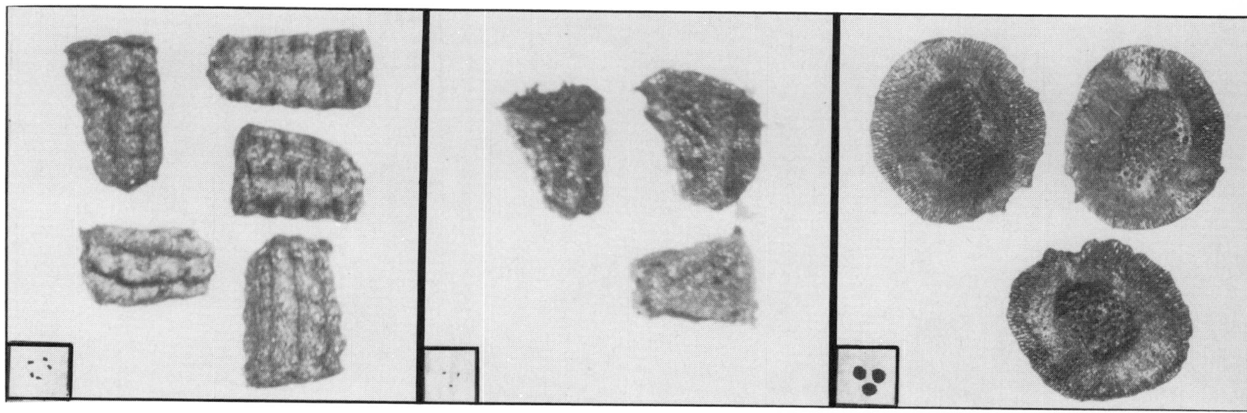

252. *Verbascum thapsus*
Mullein

253. *Linaria canadensis*
Toadflax

254. *Linaria vulgaris*
Butter-and-eggs

255. *Penstemon breviflorus*
Penstemon

256. *Castilleja sessiliflora*
Paintbrush

257. *Plantago major*
Plantain

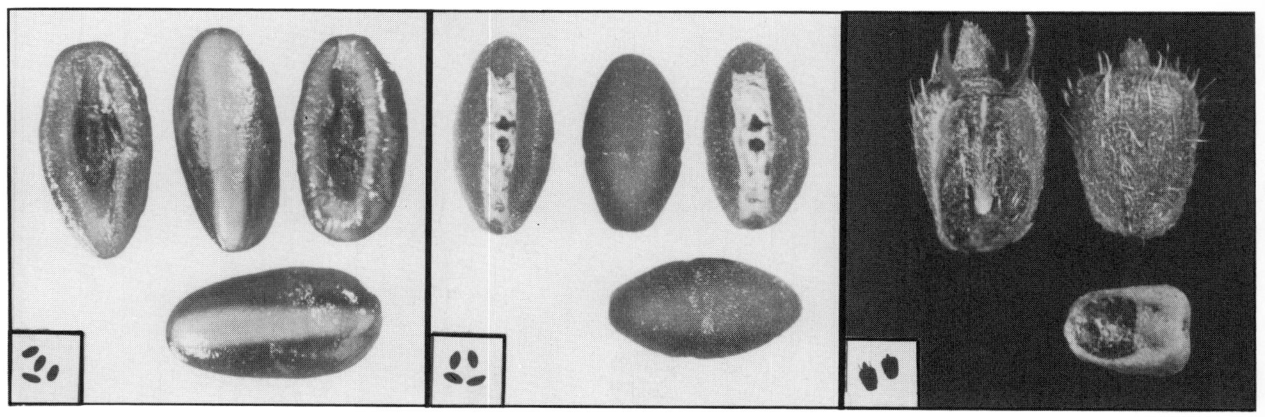

258. *Plantago lanceolata*
Buckhorn

259. *Plantago aristata*
Plantain

260. *Diodia teres*
Buttonweed

261. *Liatris squarrosa*
Gayfeather

262. *Grindelia squarrosa*
Gumweed

263. *Chrysopsis villosa*
Goldaster

264. *Gutierrezia texana*
Snakeweed

265. *Heterotheca subaxillaris*
Heterotheca

266. *Soldiago canadensis*
Goldenrod

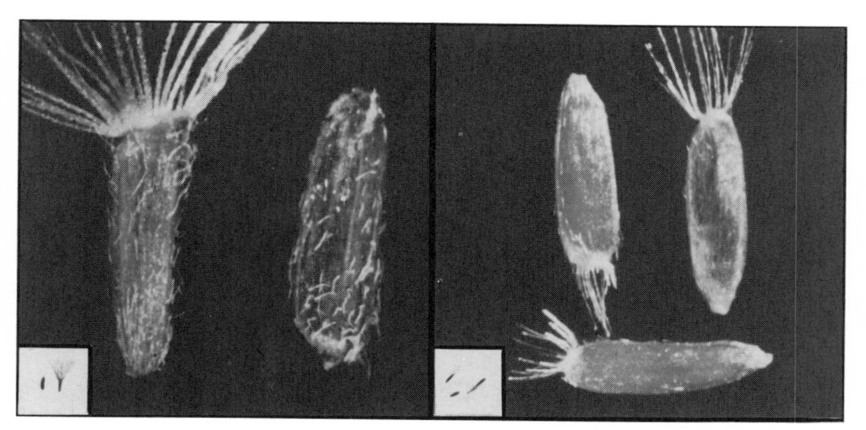

267. *Solidago serotina*
Goldenrod

268. *Erigeron canadensis*
Fleabane

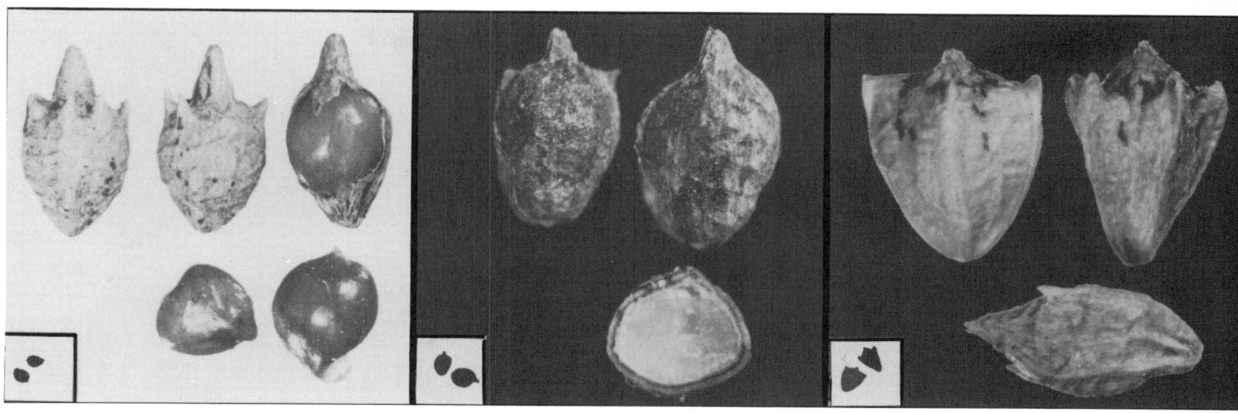

269. *Ambrosia artemisiaefolia*
Ragweed

270. *Ambrosia psilostachya*
Ragweed

271. *Ambrosia aptera*
Ragweed

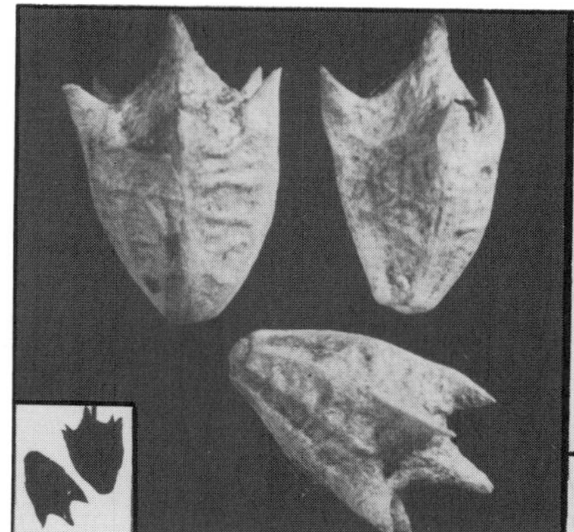

272. *Ambrosia trifida*
Ragweed

273. *Rudbeckia hirta*
Black-eyed-Susan

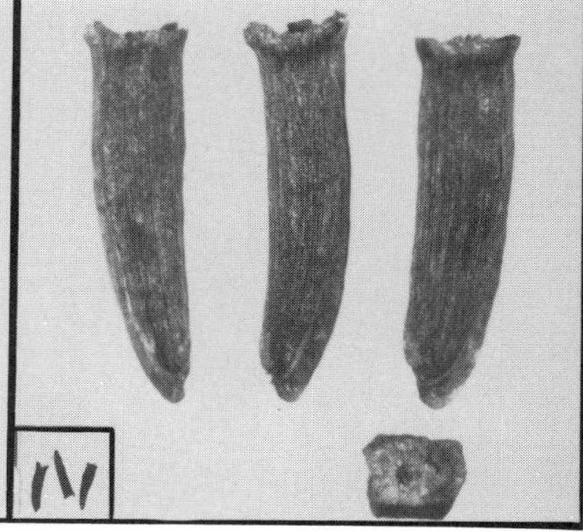

274. *Rudbeckia laciniata*
Coneflower

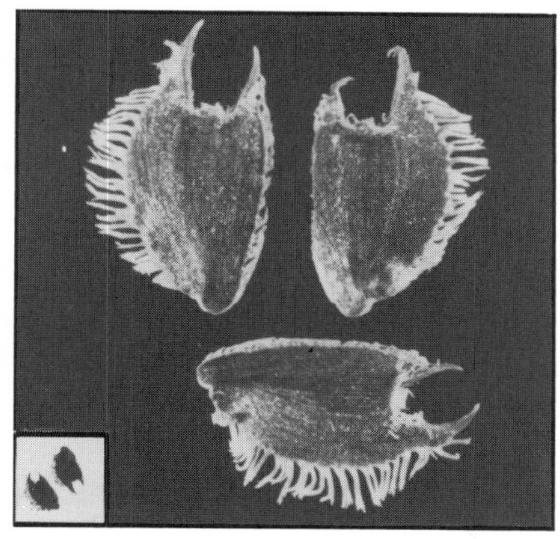

275. *Ratibida peduncularis*
Prairie-coneflower

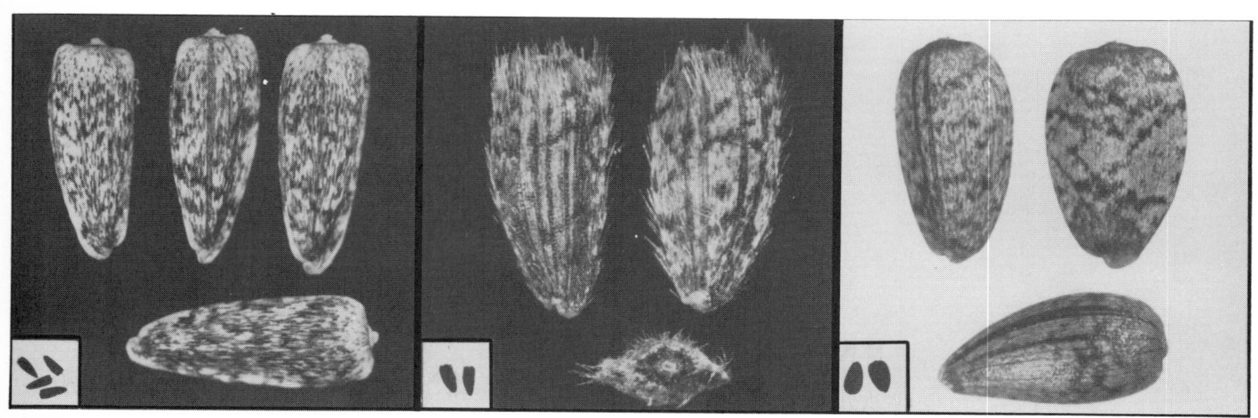

276. *Helianthus tuberosus*
Sunflower

277. *Helianthus petiolaris*
Sunflower

278. *Helianthus annuus*
Sunflower

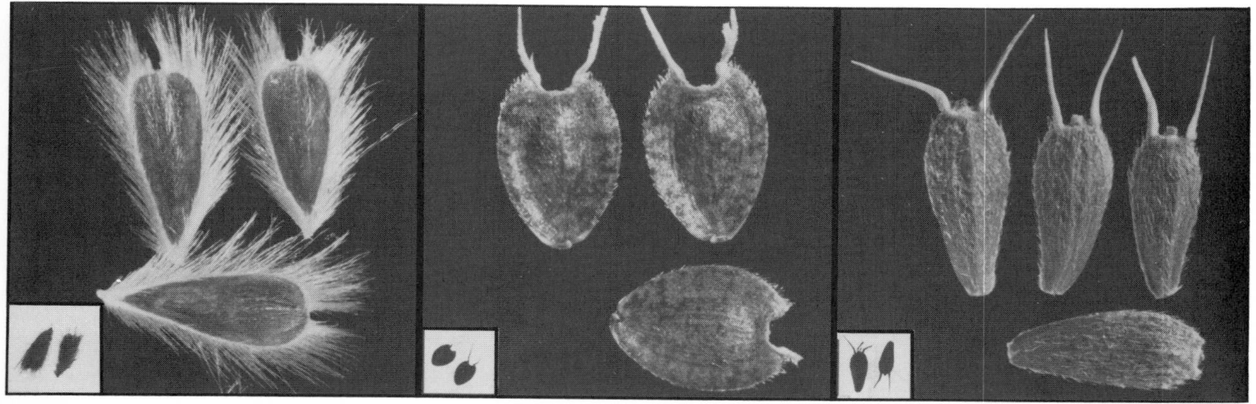

279. *Encelia frutescens*
Encelia

280. *Encelia subaristata*
Encelia

281. *Verbesina occidentalis*
Crownbeard

282. *Verbesina encelioides*
Crownbeard

283. *Verbesina alternifolia*
Crownbeard

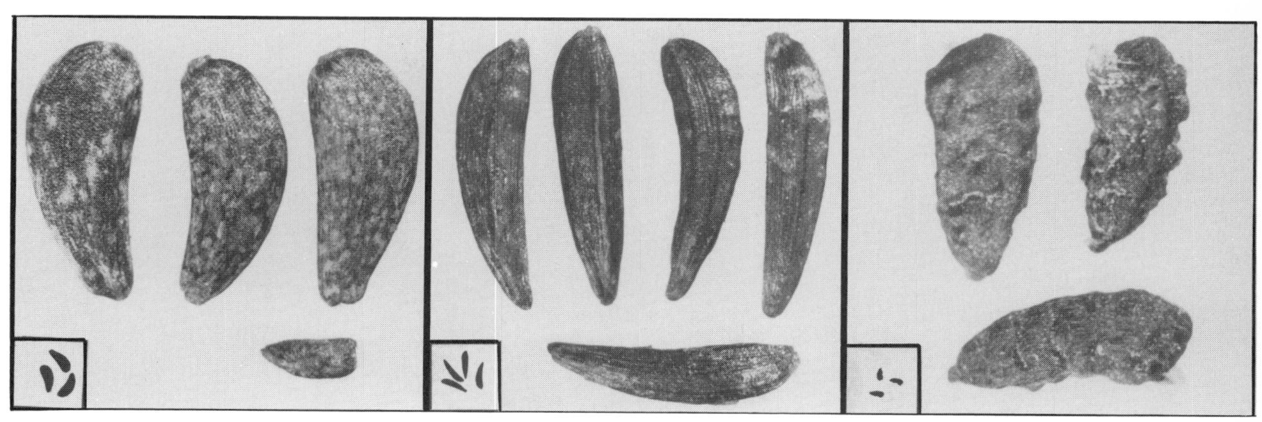

284. *Madia sativa*
Tarweed

285. *Madia glomerata*
Tarweed

286. *Hemizonia wrightii*
Tarweed

287. *Hemizonia congesta*
Tarweed

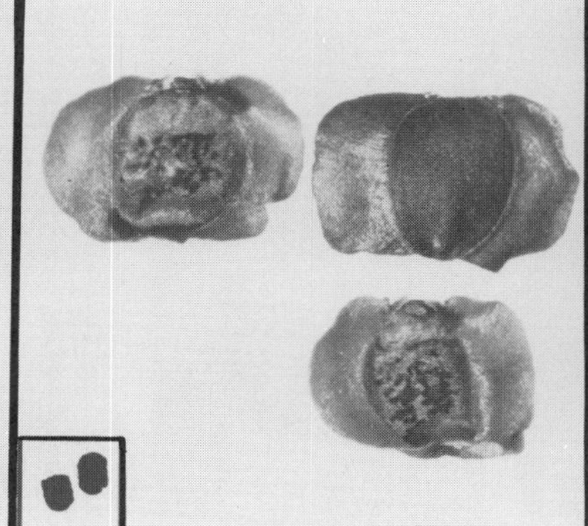

288. *Coreopsis grandiflora*
Coreopsis

289. *Gaillardia pulchella*
Gaillardia

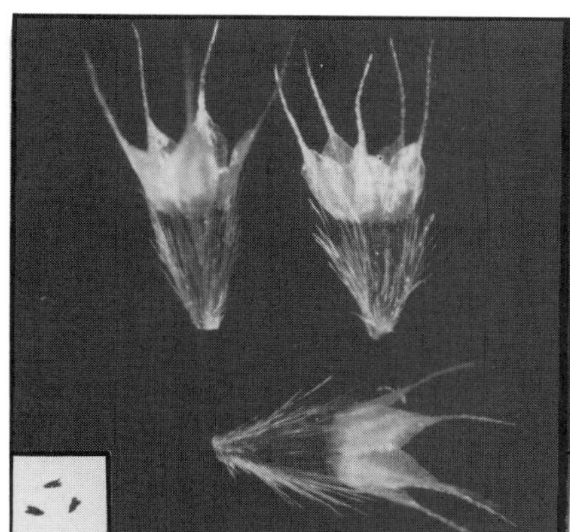

290. *Helenium tenuifolium*
Bitterweed

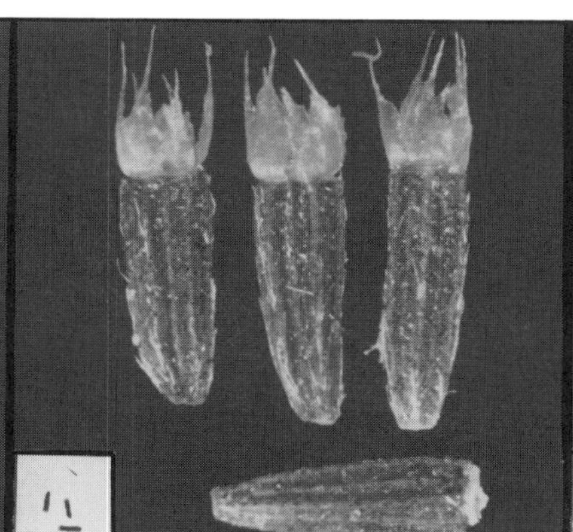

291. *Helenium puberulum*
Sneezeweed

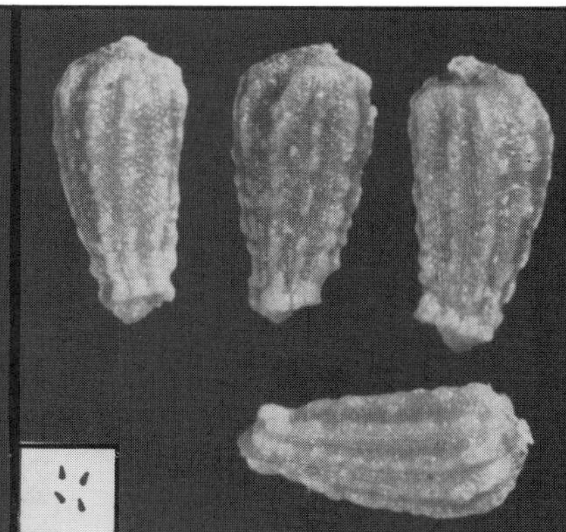

292. *Anthemis cotula*
Mayweed

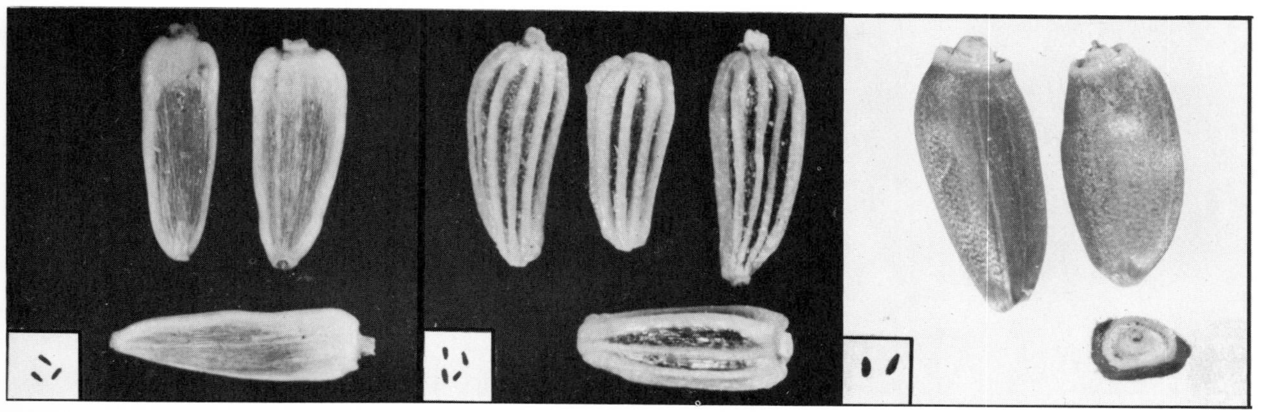

293. *Achillea millefolium*
Yarrow

294. *Chrysanthemum leucanthemum*
Oxeye Daisy

295. *Cirsium arvense*
Canada Thistle

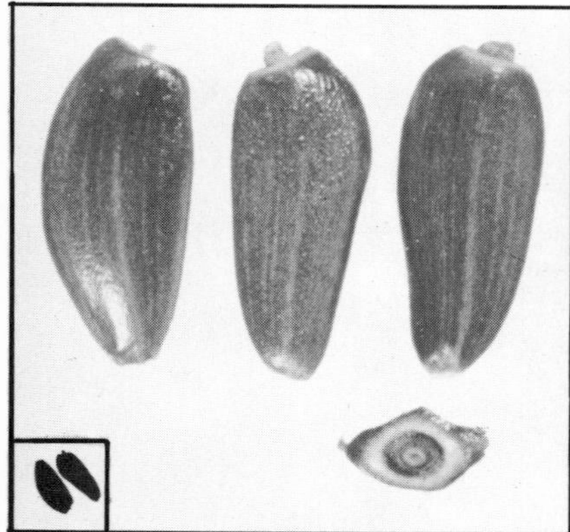

296. *Cirsium californicum*
Thistle

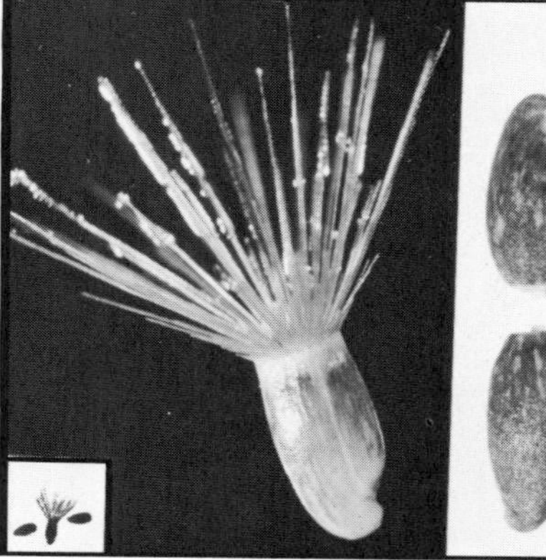

297. *Centaurea solstitialis*
Starthistle

298. *Centaurea melitensis*
Centaurea

299. *Centaurea americana*
Centaurea

300. *Silybum marianum*
Milk Thistle

301. *Cichorium intybus*
Chicory

302. *Serinea oppositifolia*
Serinea

303. *Taraxacum officinale*
Dandelion

304. *Sonchus asper*
Sowthistle

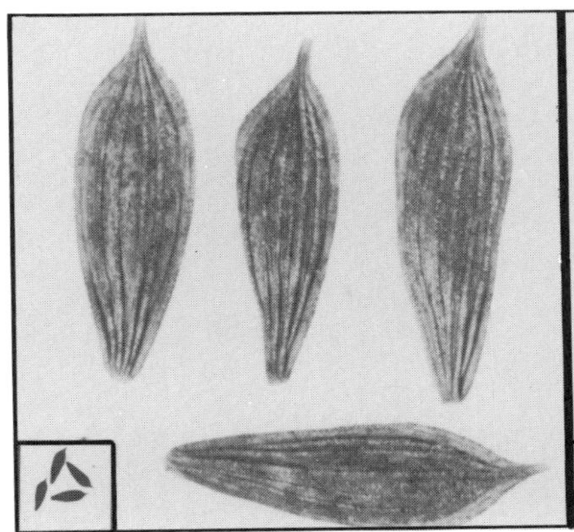

305. *Lactuca scariola*
Wild-lettuce

306. *Lactuca canadensis*
Wild-lettuce

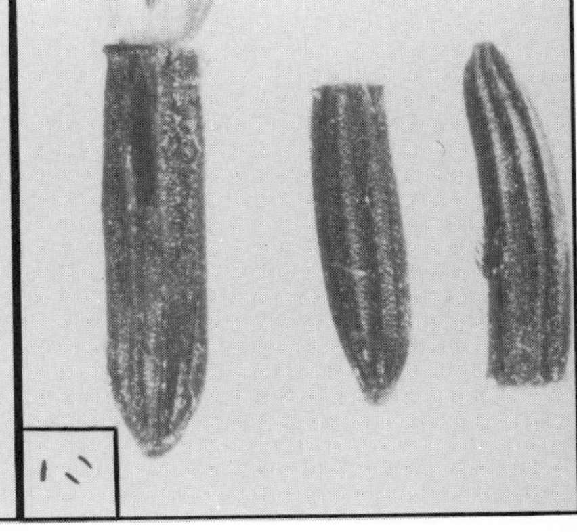

307. *Hieracium aurantiacum*
Hawkweed

Farmlands

Weeds, Arranged by Physical Features
(Weeds II)

Because of the importance of weeds to investigators in various fields, it has seemed worthwhile to present a second gallery of weed seeds, arranged according to their prominent external characteristics. This arrangement will help one to locate seeds rapidly, without hunting through the whole systematic sequence of pictures. Many of the same seeds shown in natural-family order in the preceding subdivision are here displayed again, classified by shape or by similar kinds of attachments, such as wings, pappus, or awns. The illustrated key below (fig. 1) indicates the physical criteria used and the plates included in each group.

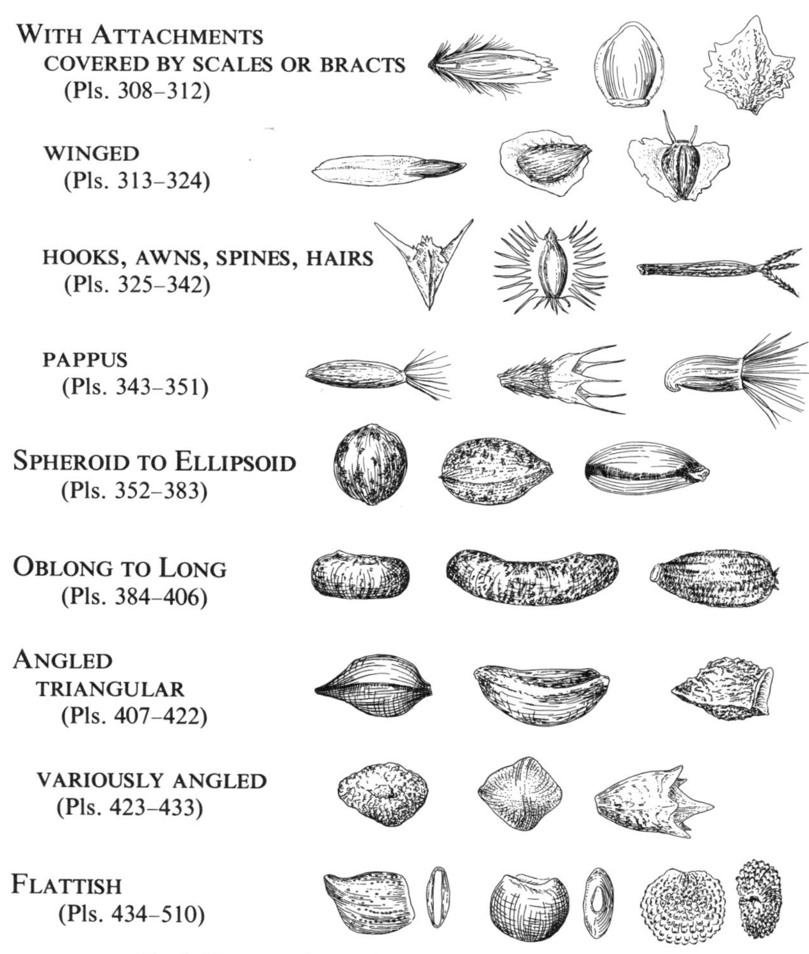

Fig. 1. Key to seed types, based on attachments and shapes

WEEDS II WITH ATTACHMENTS

PLATES 308–315

COVERED BY SCALES OR BRACTS

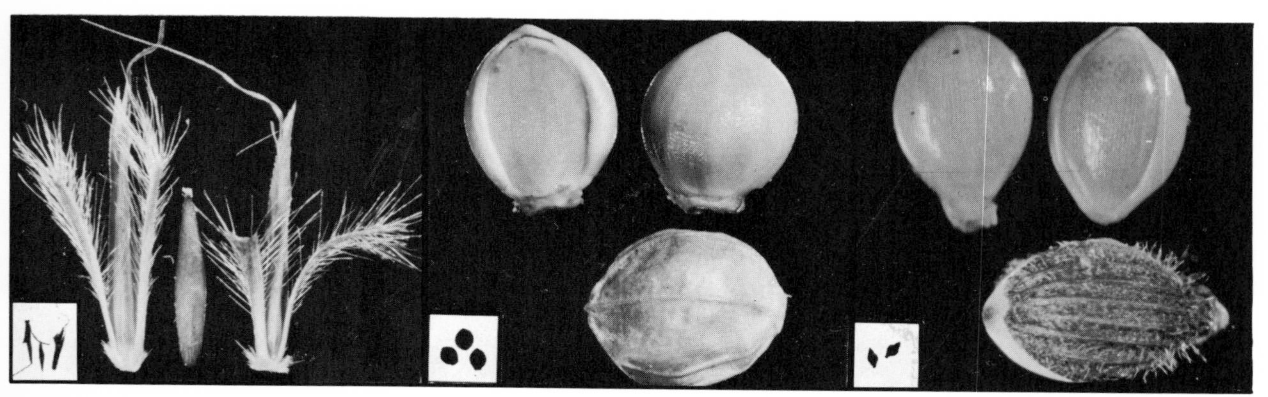

308. *Andropogon scoparius*
Bluestem

309. *Paspalum laeve*
Paspalum

310. *Panicum lindheimeri*
Panicum

Additional scale-covered seeds are pictured in plates 1–44 and figure 24

311. *Digitaria ischaemum*
Crabgrass

312. *Atriplex semibaccata*
Saltbush

WINGED

313. *Oenothera biennis*
Evening-primrose

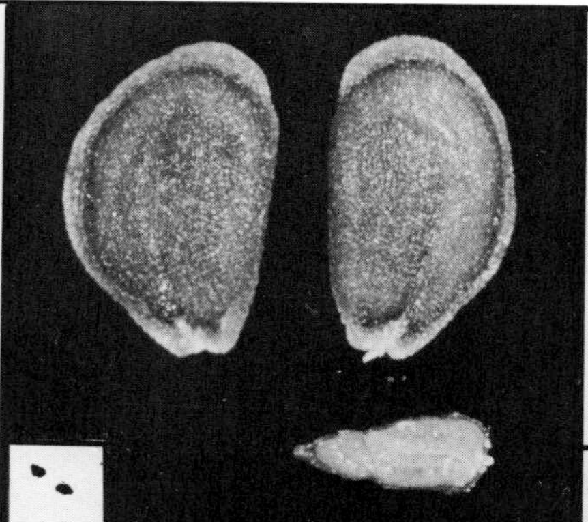

314. *Lepidium virginicum*
Pepperweed

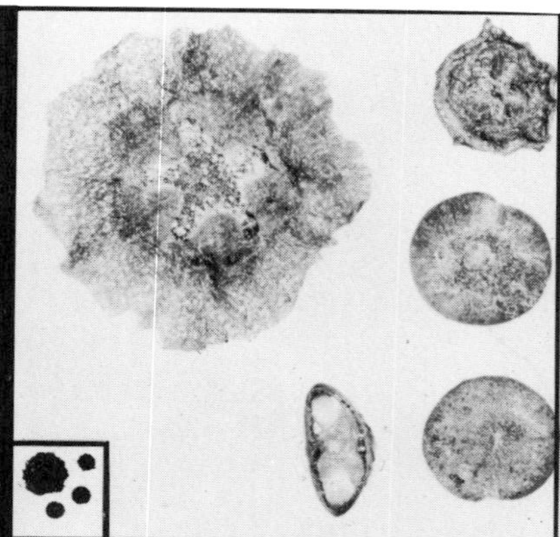

315. *Cycloloma atriplicifolium*
Ringwing

WINGED

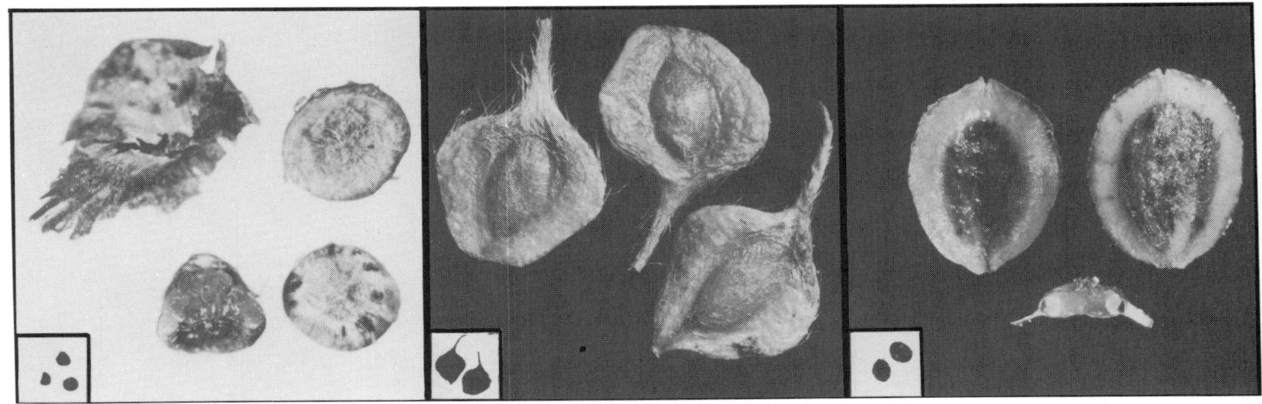

316. *Salsola kali*
Russianthistle

317. *Anemone canadensis*
Anemone

318. *Corispermum hyssopifolium*
Tickseed

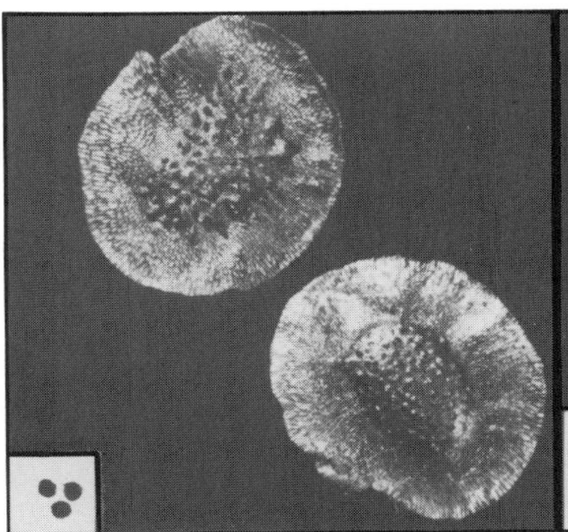

319. *Linaria vulgaris*
Butter-and-eggs

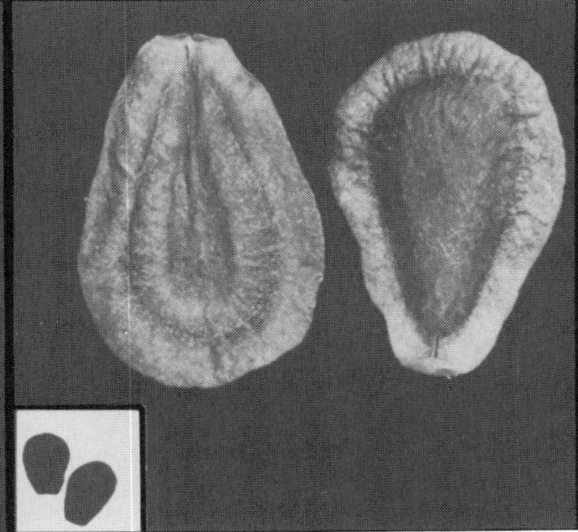

320. *Asclepias syriaca*
Milkweed

321. *Mentzelia decapetala*
Prairiestar

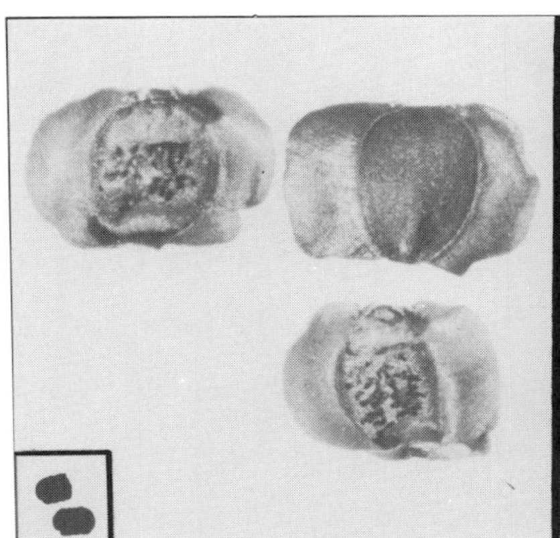

322. *Coreopsis grandiflora*
Coreopsis

323. *Verbesina encelioides*
Crownbeard

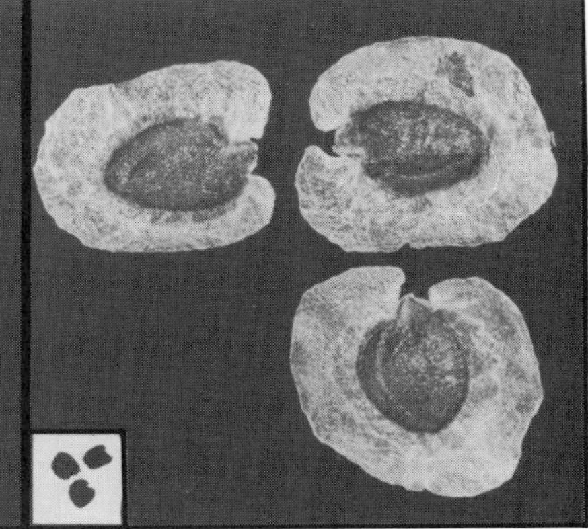

324. *Arabis canadensis*
Rockcress

WEEDS II WITH ATTACHMENTS
HOOKS, AWNS, SPINES, HAIRS

PLATES 325–333

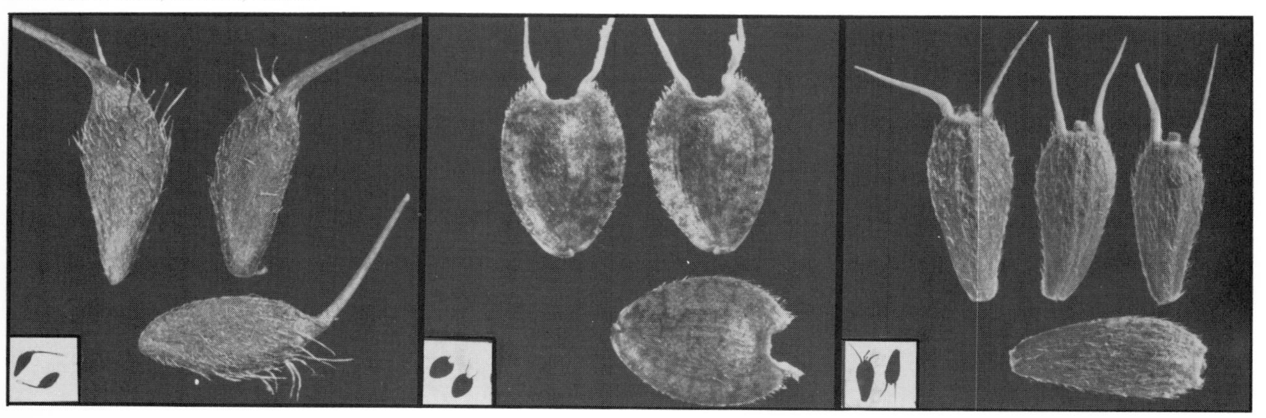

325. *Geum canadense*
Avens

326. *Encelia subaristata*
Encelia

327. *Verbesina occidentalis*
Crownbeard

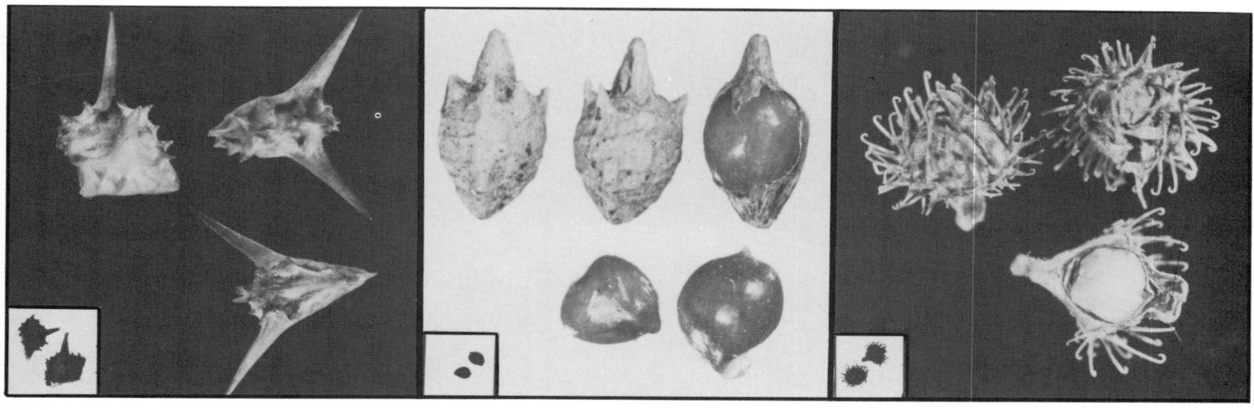

328. *Tribulus terrestris*
Puncturevine

329. *Ambrosia artemisiaefolia*
Ragweed

330. *Agrimonia parviflora*
Agrimony

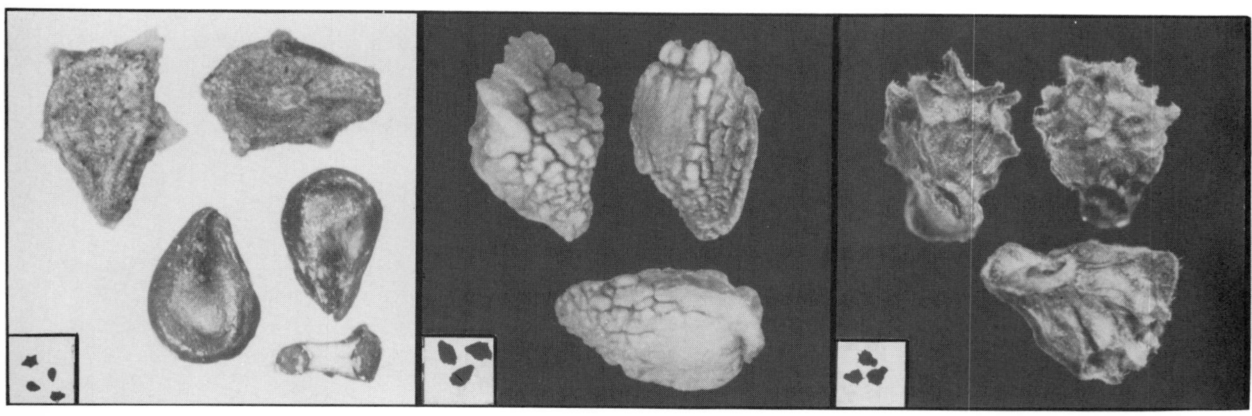

331. *Kochia scoparia*
Summer-cypress

332. *Oenothera missouriensis*
Sundrops

333. *Trichostema lanceolatum*
Bluecurls

WEEDS II WITH ATTACHMENTS PLATES 334–342

HOOKS, AWNS, SPINES, HAIRS

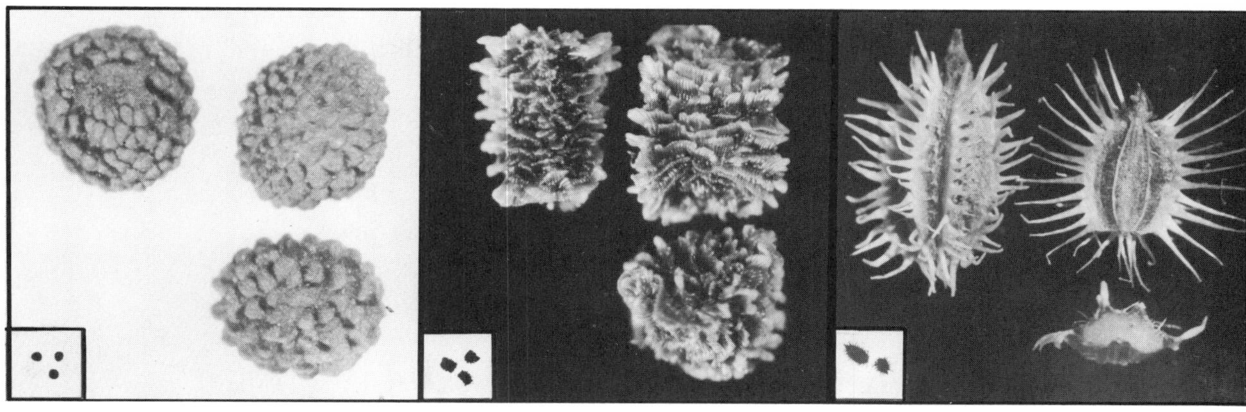

334. *Scutellaria integrifolia*
Skullcap

335. *Delphinium virescens*
Larkspur

336. *Daucus carota*
Wildcarrot

337. *Ratibida peduncularis*
Prairie-coneflower

338. *Encelia frutescens*
Encelia

339. *Ipomoea pandurata*
Morning-glory

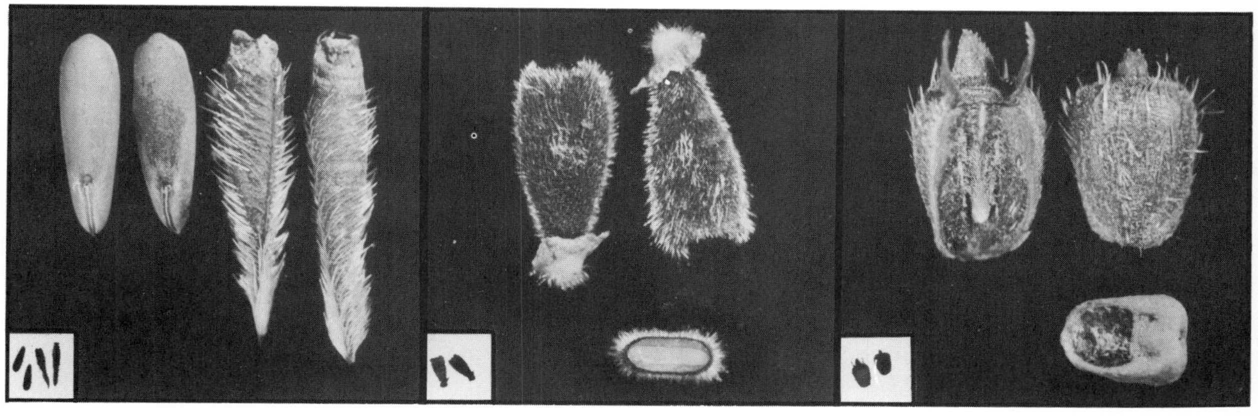

340. *Erodium botrys*
Filaree

341. *Polygala longa*
Milkwort

342. *Diodia teres*
Buttonweed

WEEDS II WITH ATTACHMENTS PLATES 343–351

PAPPUS

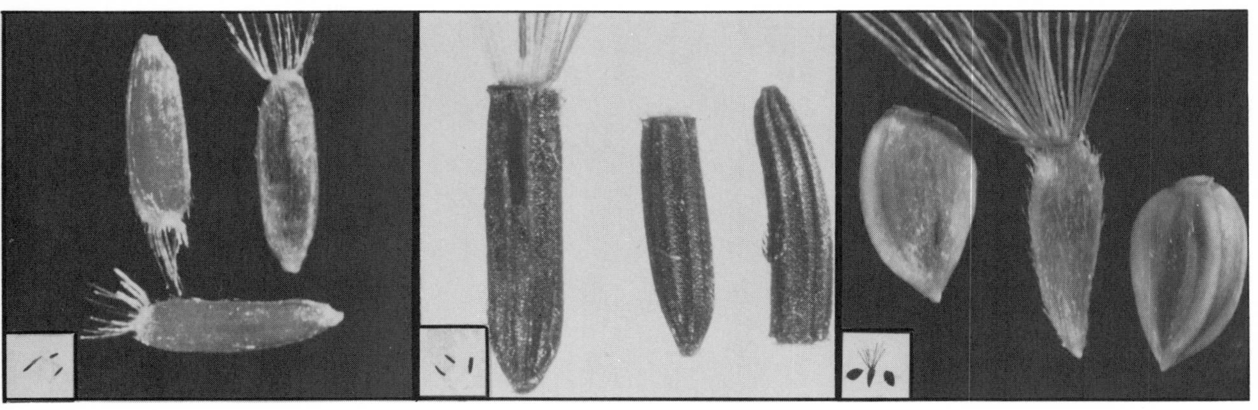

343. *Erigeron canadensis*
Fleabane

344. *Hieracium aurantiacum*
Hawkweed

345. *Heterotheca subaxillaris*
Heterotheca

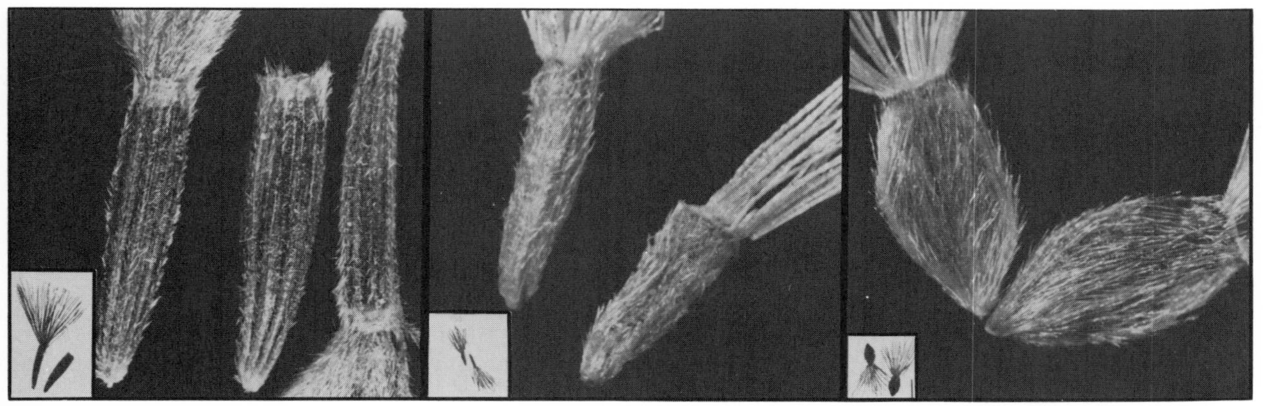

346. *Liatris squarrosa*
Gayfeather

347. *Solidago canadensis*
Goldenrod

348. *Chrysopsis villosa*
Goldaster

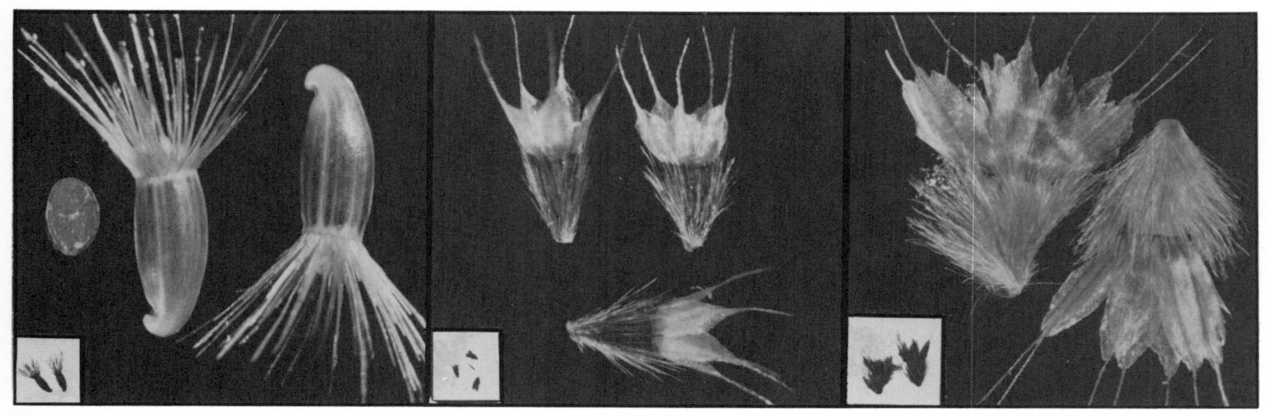

349. *Centaurea melitensis*
Centaurea

350. *Helenium tenuifolium*
Bitterweed

351. *Gaillardia pulchella*
Gaillardia

352. *Tragia urens*
Noseburn

353. *Lotus tomentellus*
Deervetch

354. *Spergula arvensis*
Spurry

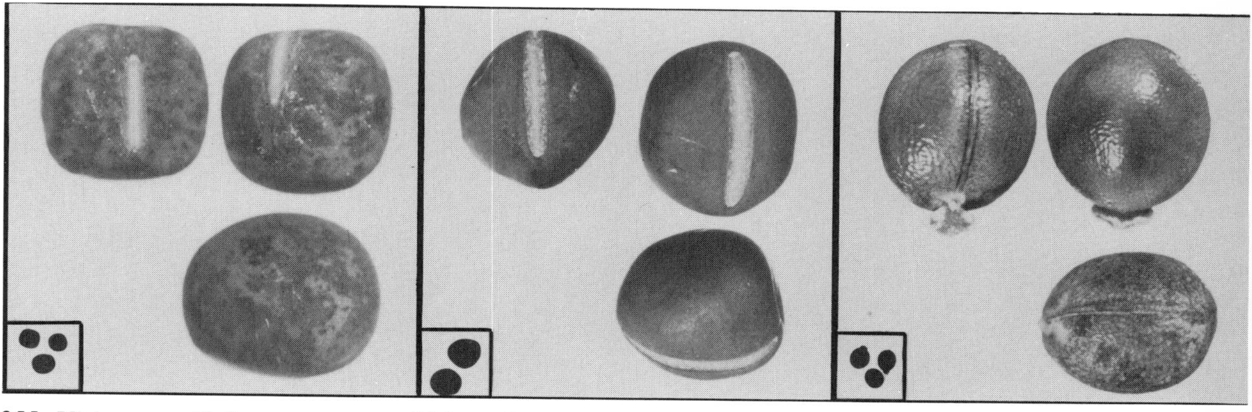

355. *Vicia angustifolia*
Vetch

356. *Lathyrus maritimus*
Peavine

357. *Croton monanthogynus*
Doveweed

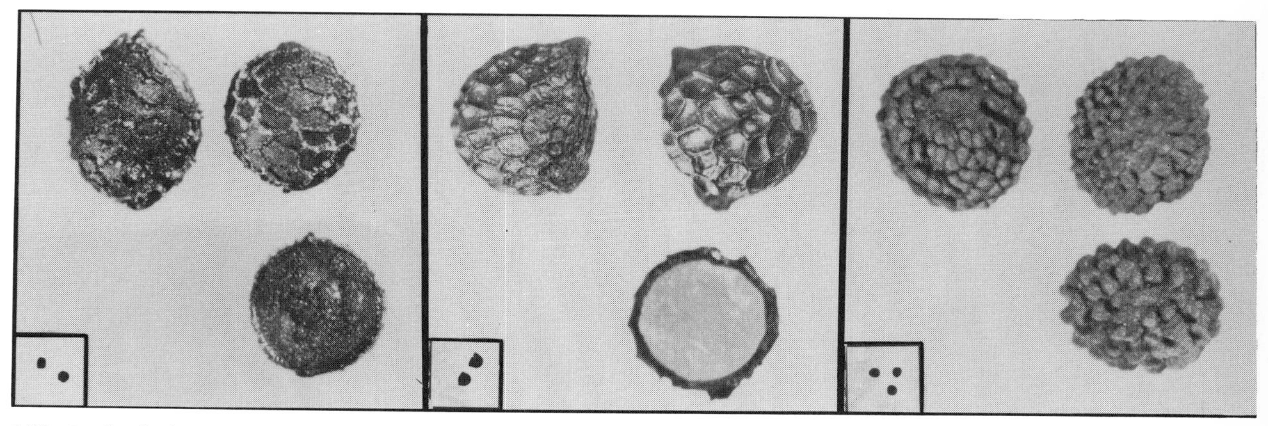

358. *Eschscholtzia californica*
California-poppy

359. *Argemone platyceras*
Pricklypoppy

360. *Scutellaria integrifolia*
Skullcap

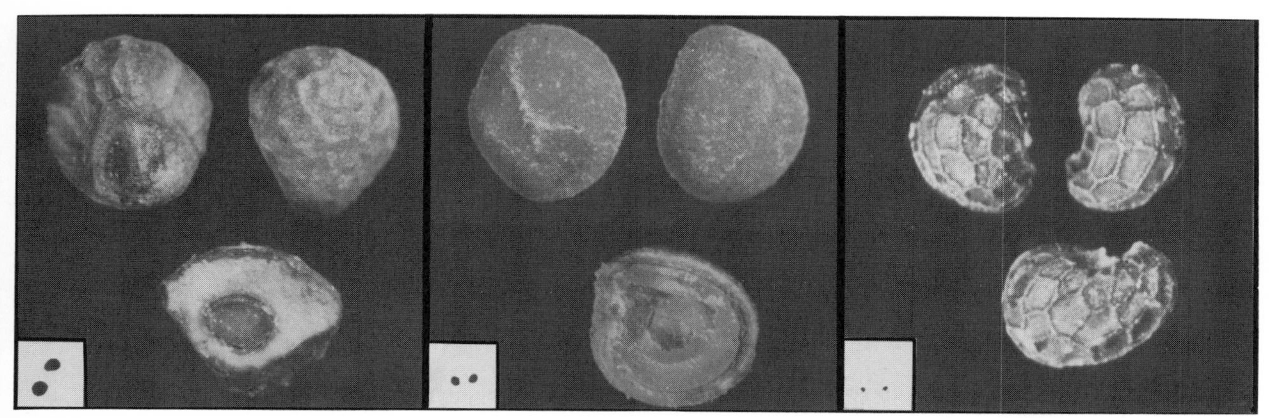

361. *Teucrium canadense*
Germander

362. *Cuscuta gronovii*
Dodder

363. *Papaver dubium*
Poppy

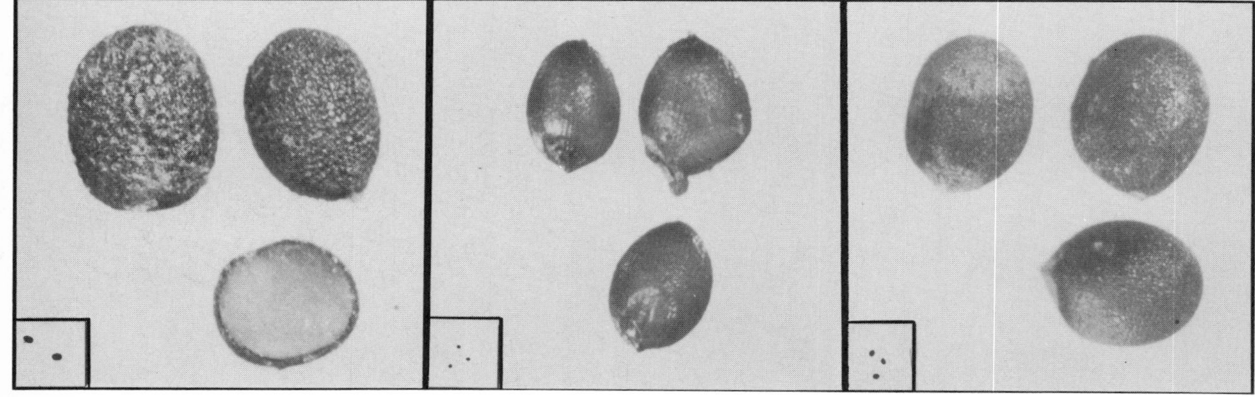

364. *Brassica nigra*
Mustard

365. *Eragrostis cilianensis*
Stinkgrass

366. *Hedeoma pulegioides*
False Pennyroyal

367. *Euphorbia corollata*
Spurge

368. *Acalypha virginica*
Copperleaf

369. *Lepidium campestre*
Pepperweed

370. *Geranium carolinianum*
Geranium

371. *Strophostyles umbellata*
Wildbean

372. *Schrankia angustata*
Sensitivebrier

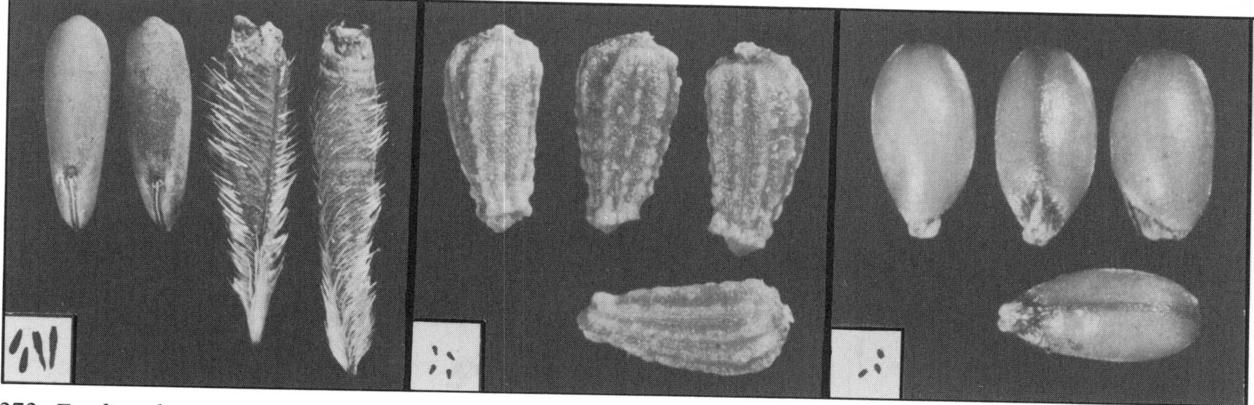

373. *Erodium botrys*
Filaree

374. *Anthemis cotula*
Mayweed

375. *Viola refinesquii*
Johnny-jump-up

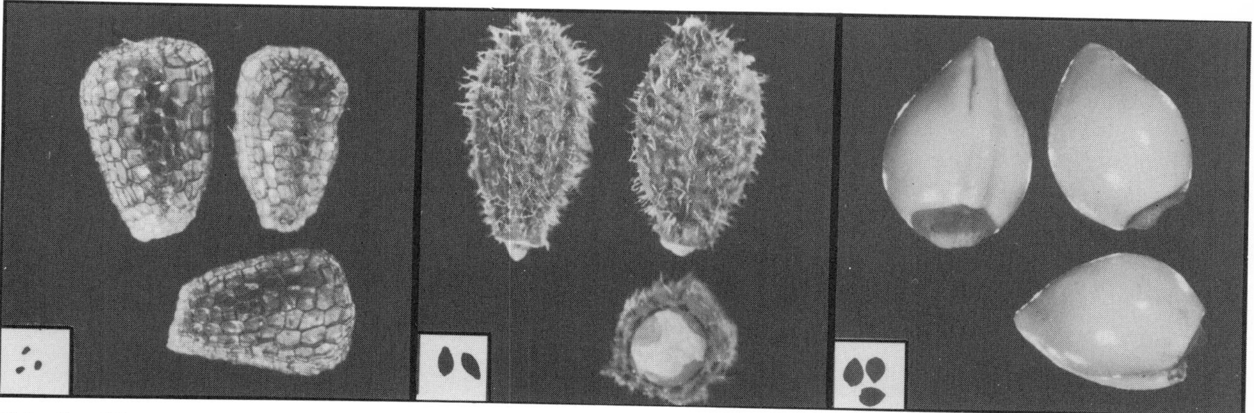

376. *Castilleja sessiliflora*
Paintbrush

377. *Oxybaphus nyctagineus*
Umbrellawort

378. *Lithospermum carolinense*
Gromwell

379. *Echinocactus rigidissimus*
Cactus

380. *Salsola kali*
Russianthistle

381. *Commelina crispa*
Dayflower

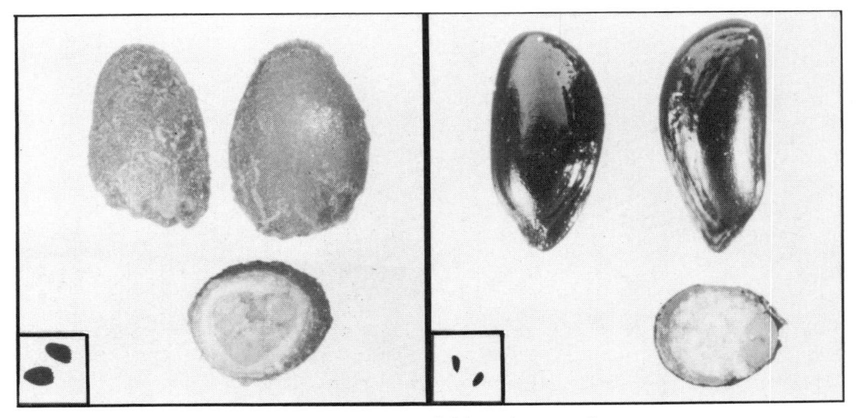

382. *Breweria pickeringii*
Breweria

383. *Abronia fragrans*
Sandverbena

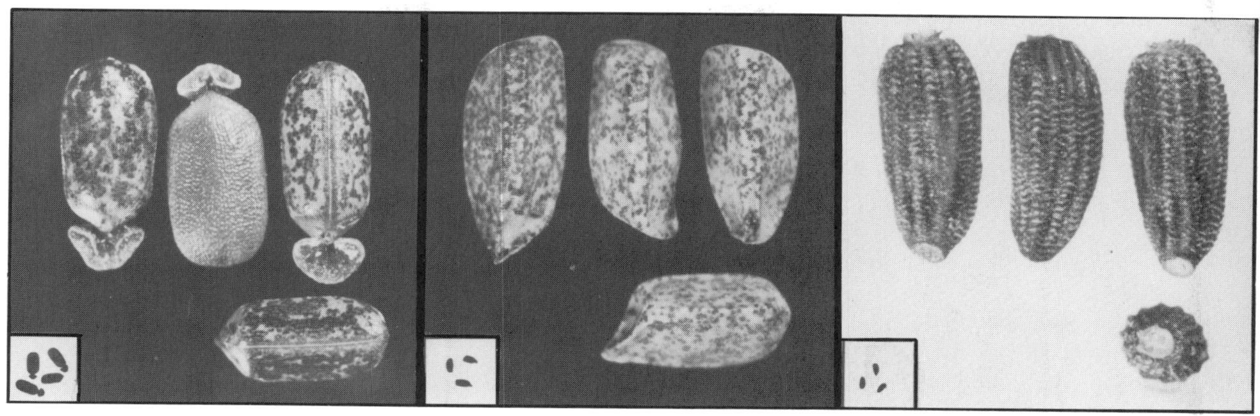

384. *Croton lindheimerianus*
Doveweed

385. *Mentzelia micrantha*
Prairiestar

386. *Serinea oppositifolia*
Serinea

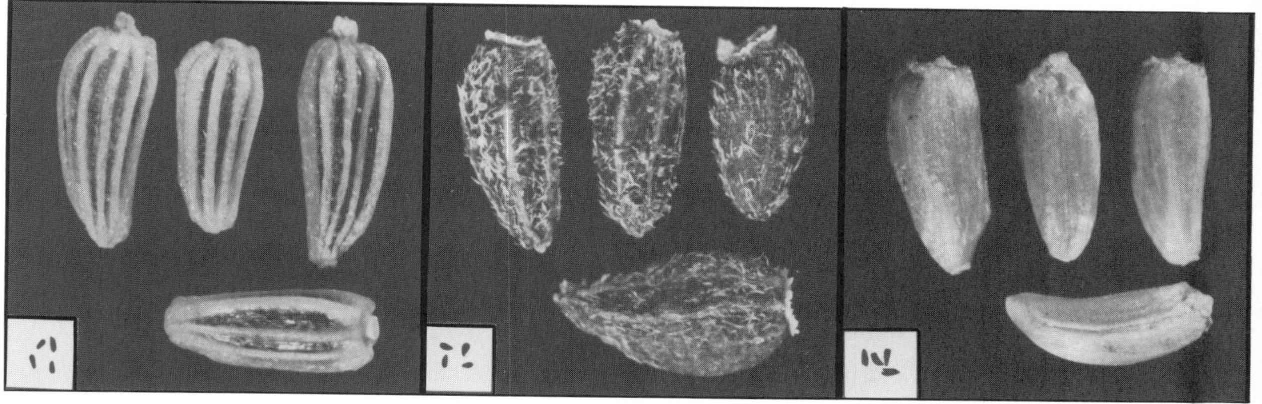

387. *Chrysanthemum leucanthemum*
Oxeye Daisy

388. *Gutierrezia texana*
Snakeweed

389. *Grindelia squarrosa*
Gumweed

390. *Sesbania macrocarpa*
Coffeeweed

391. *Cichorium intybus*
Chicory

392. *Cirsium arvense*
Canada Thistle

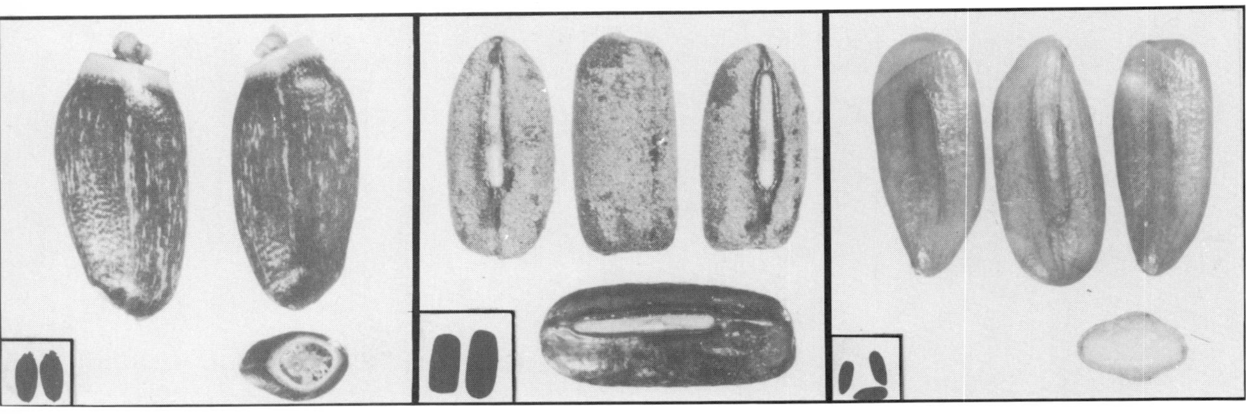

393. *Silybum marianum*
Milk Thistle

394. *Strophostyles helvola*
Wildbean

395. *Desmanthus leptolobus*
Bundleflower

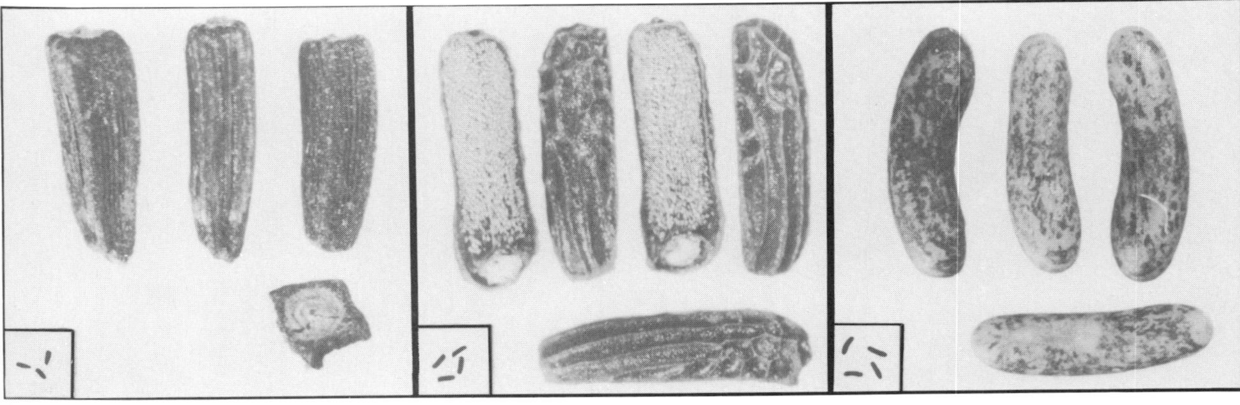

396. *Rudbeckia hirta*
Black-eyed-Susan

397. *Verbena bracteosa*
Verbena

398. *Lotus scoparius*
Deervetch

399. *Taraxacum officinale*
Dandelion

400. *Erodium cicutarium*
Filaree

401. *Helianthus tuberosus*
Sunflower

402. *Berberis vulgaris*
Barberry

403. *Eriogonum alatum*
Eriogonum

404. *Lactuca scariola*
Wild-lettuce

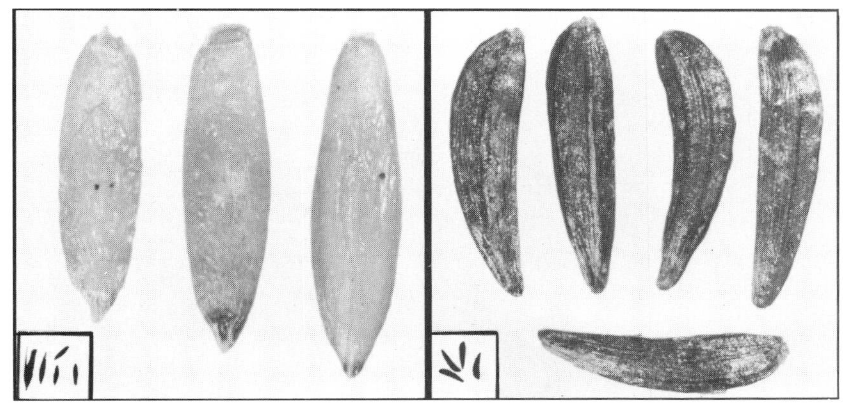

405. *Bouteloua curtipendula*
Grama

406. *Madia glomerata*
Tarweed

WEEDS II ANGLED
TRIANGULAR

PLATES 407–415

407. *Berberis aquifolium*
Oregongrape

408. *Eremocarpus setigerus*
Turkeymullein

409. *Sida spinosa*
Sida

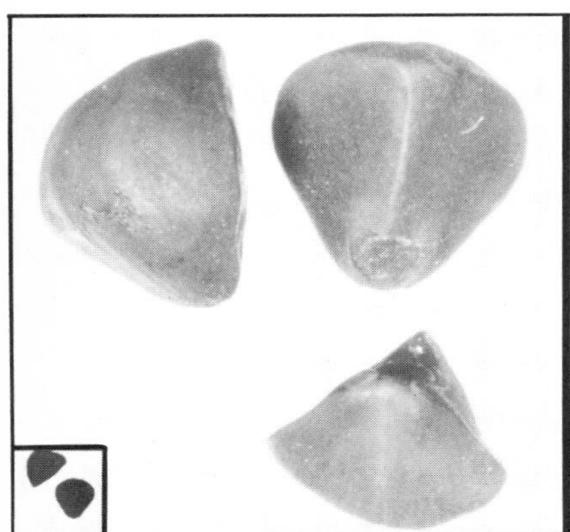

410. *Ipomoea lacunosa*
Morning-glory

411. *Convolvulus arvensis*
Bindweed

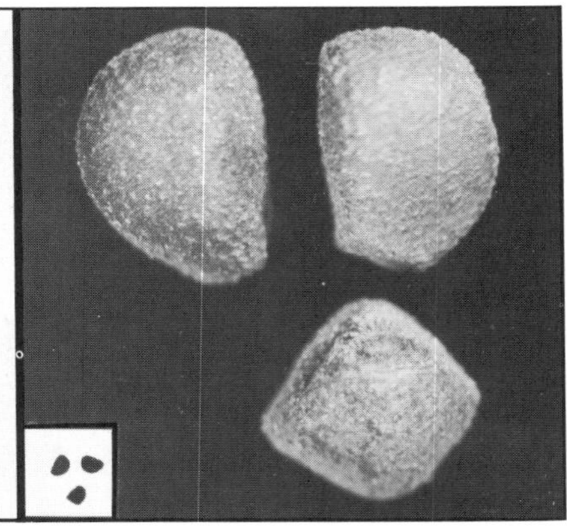

412. *Jacquemontia tamnifolia*
Jacquemontia

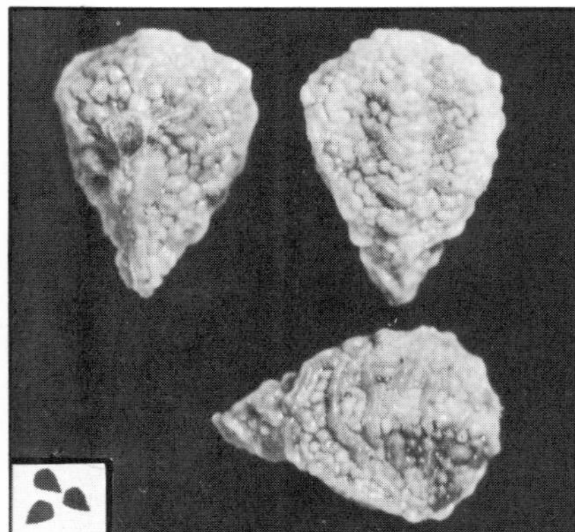

413. *Amsinckia tessellata*
Fiddleneck

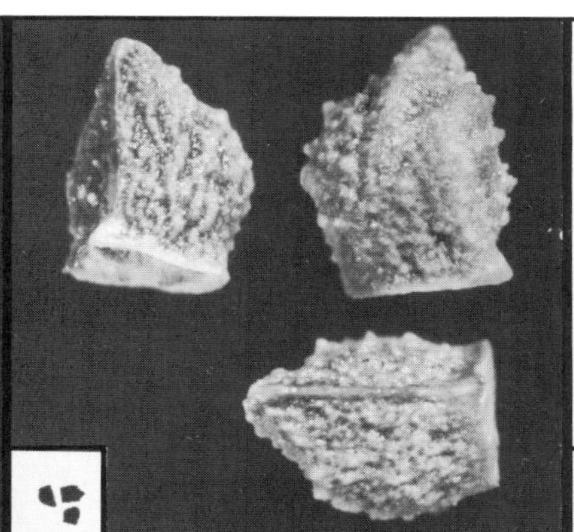

414. *Echium vulgare*
Blueweed

415. *Hemizonia congesta*
Tarweed

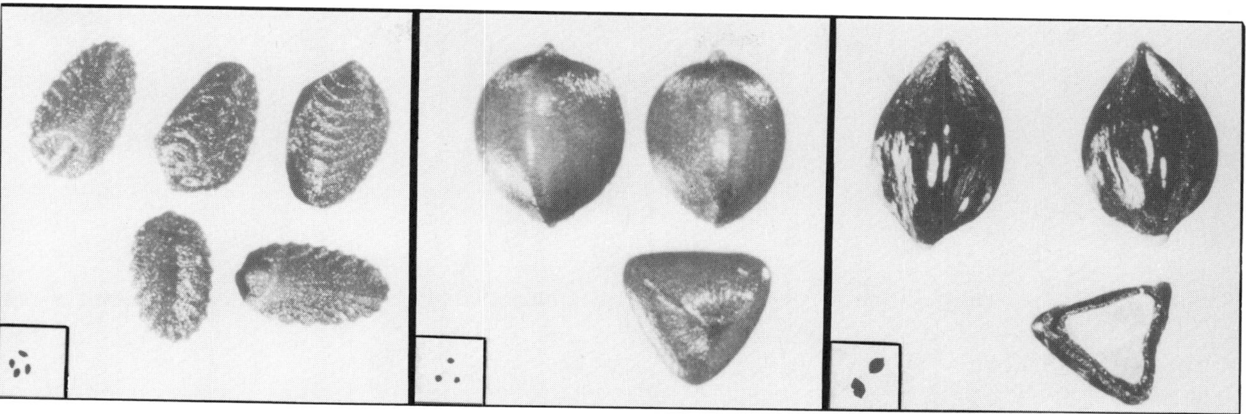

416. *Eleusine indica*
Goosegrass

417. *Rumex acetosella*
Sheep Sorrel

418. *Polygonum cilinode*
Knotweed

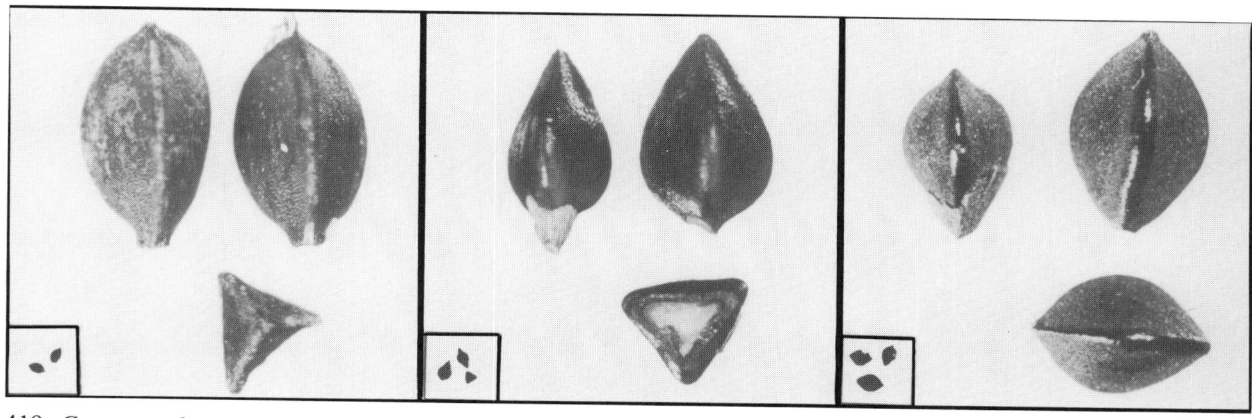

419. *Cyperus schweinitzii*
Flatsedge

420. *Polygonum aviculare*
Knotweed

421. *Polygonum convolvulus*
Cornbind

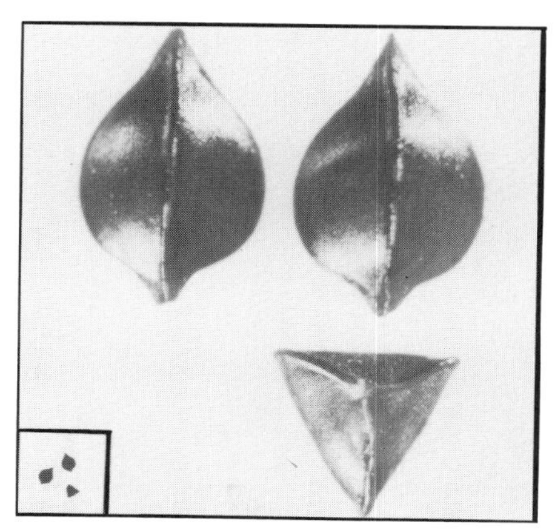

422. *Rumex crispus*
Dock

See also species of *Polygonum* illustrated in plates 53–55, 631–642, and figures 71–76

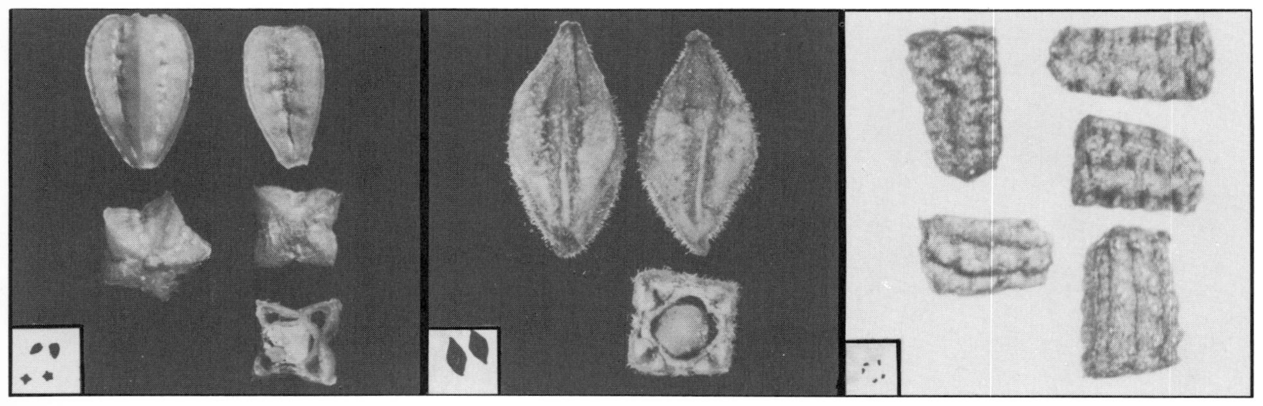

423. *Boerhaavia wrightii*
Spiderling

424. *Gaura biennis*
Gaura

425. *Verbascum thapsus*
Mullein

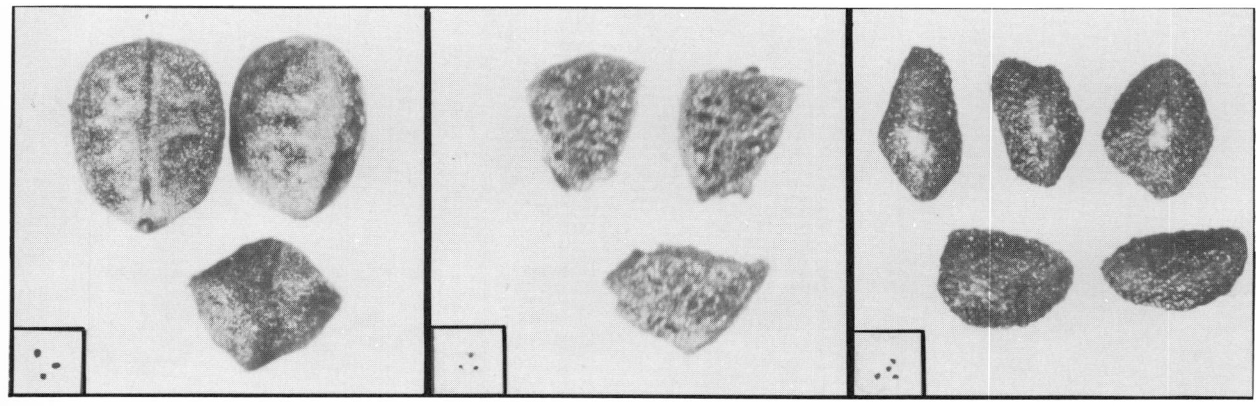

426. *Euphorbia maculata*
Spurge

427. *Linaria candensis*
Toadflax

428. *Plantago major*
Plantain

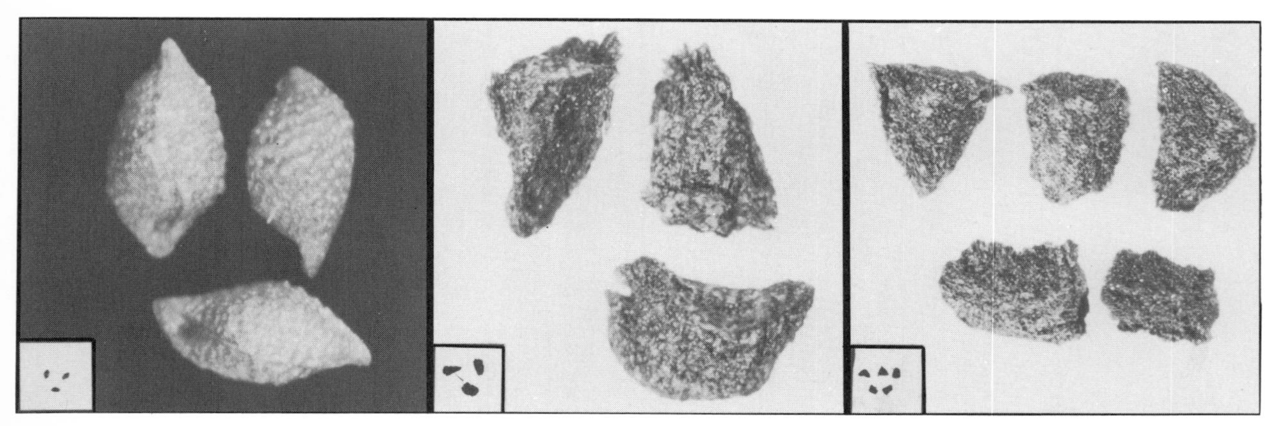

429. *Oenothera laciniata*
Evening-primrose

430. *Delphinium cardinale*
Larkspur

431. *Penstemon breviflorus*
Penstemon

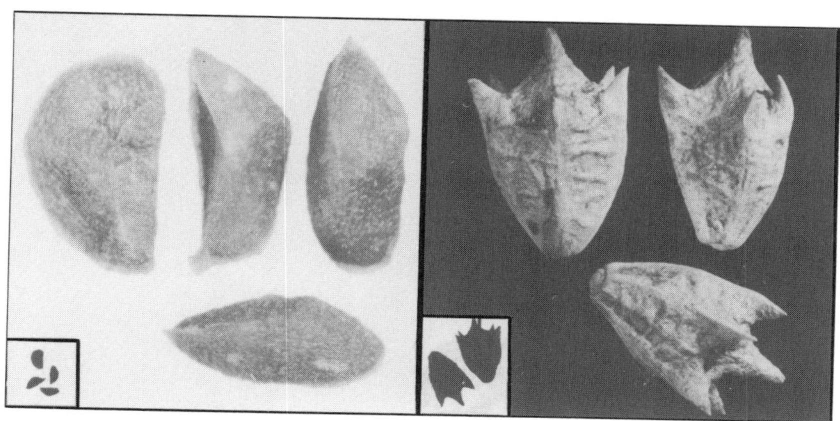

432. *Gilia aggregata*
Gilia

433. *Ambrosia trifida*
Ragweed

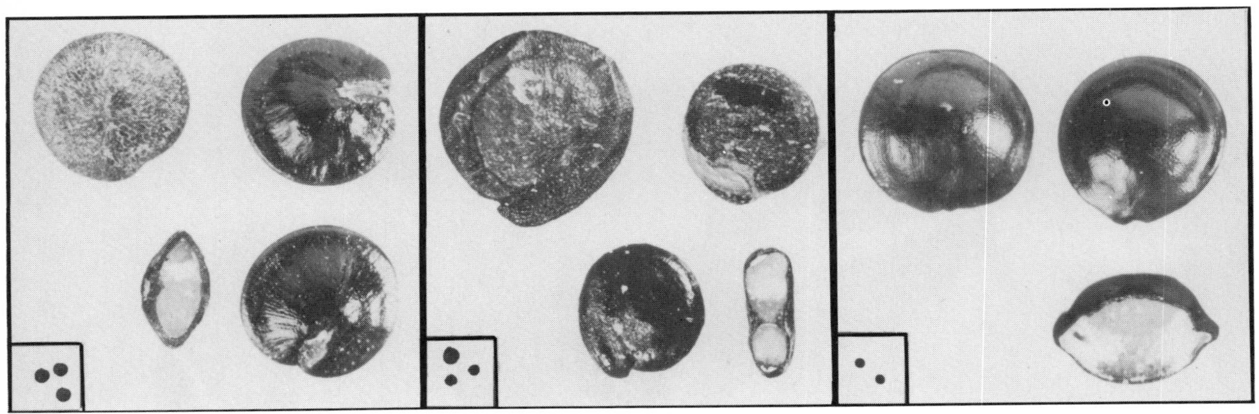

434. *Chenopodium hybridum*
Goosefoot

435. *Atriplex patula*
Saltbush

436. *Amaranthus blitoides*
Pigweed

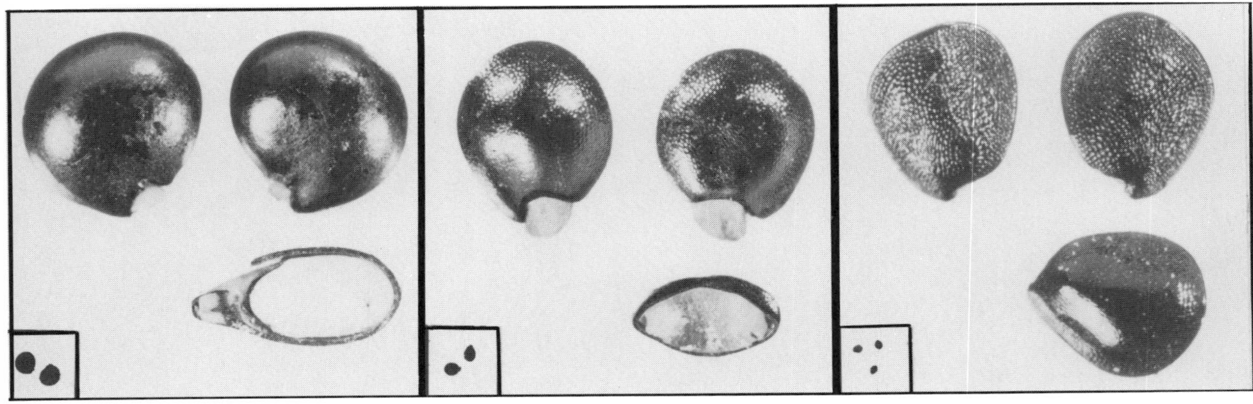

437. *Phytolacca americana*
Pokeberry

438. *Montia perfoliata*
Minerslettuce

439. *Calandrinia caulescens*
Redmaids

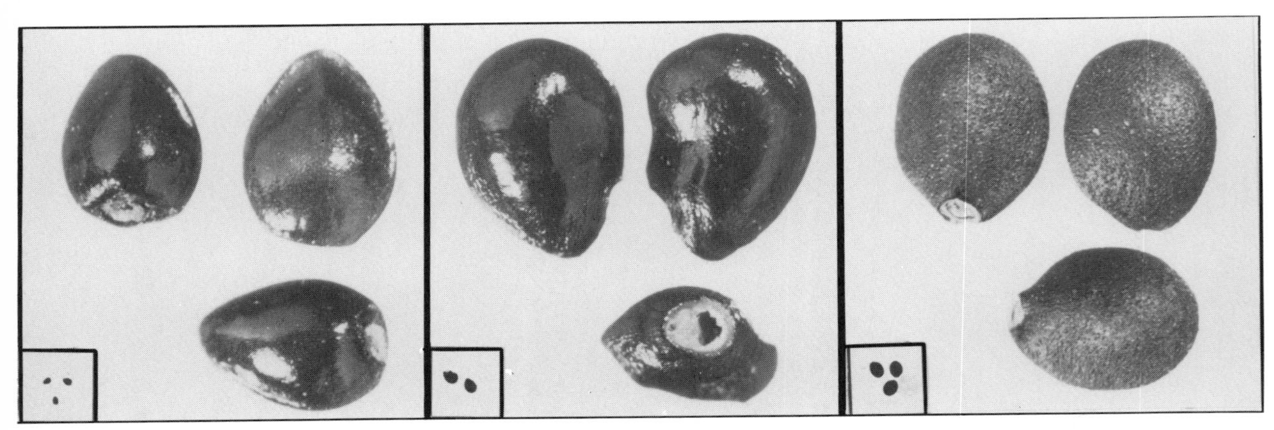

440. *Myosotis arvensis*
Forget-me-not

441. *Cereus giganteus*
Giantcactus

442. *Salvia lyrata*
Sage

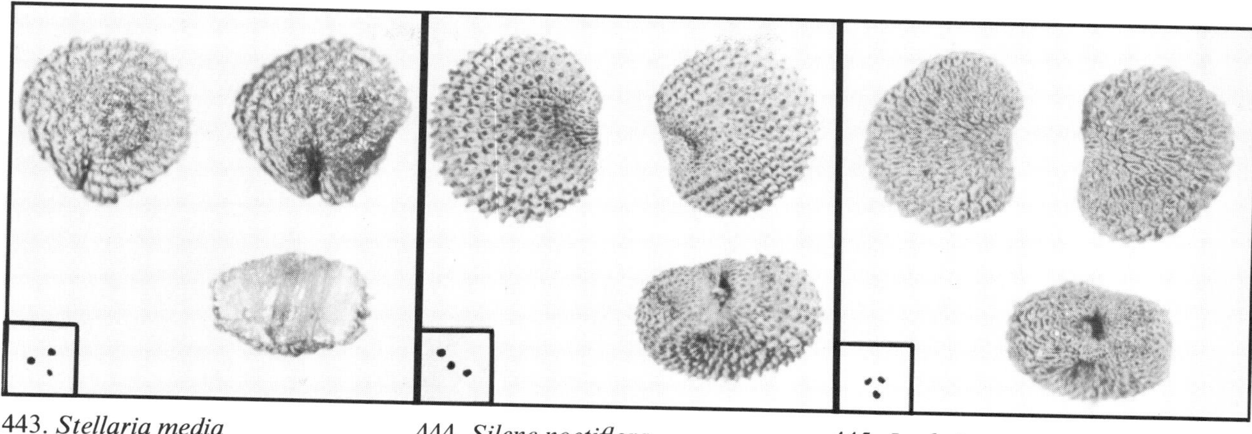

443. *Stellaria media*
Chickweed

444. *Silene noctiflora*
Catchfly

445. *Lychnis coronaria*
Campion

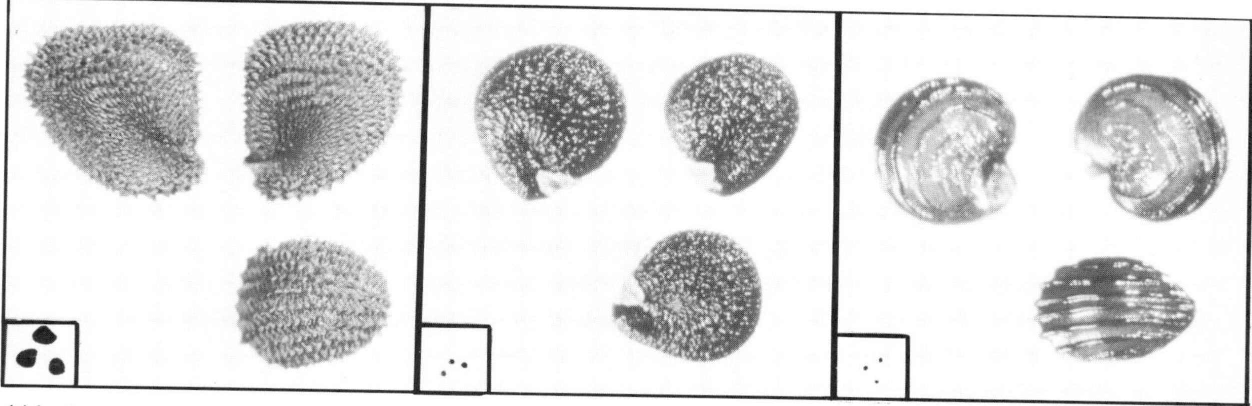

446. *Agrostemma githago*
Cockle

447. *Portulaca oleracea*
Purslane

448. *Mollugo verticillata*
Carpetweed

449. *Euphorbia dictyosperma*
Spurge

450. *Passiflora incarnata*
Passionflower

451. *Oxalis stricta*
Oxalis

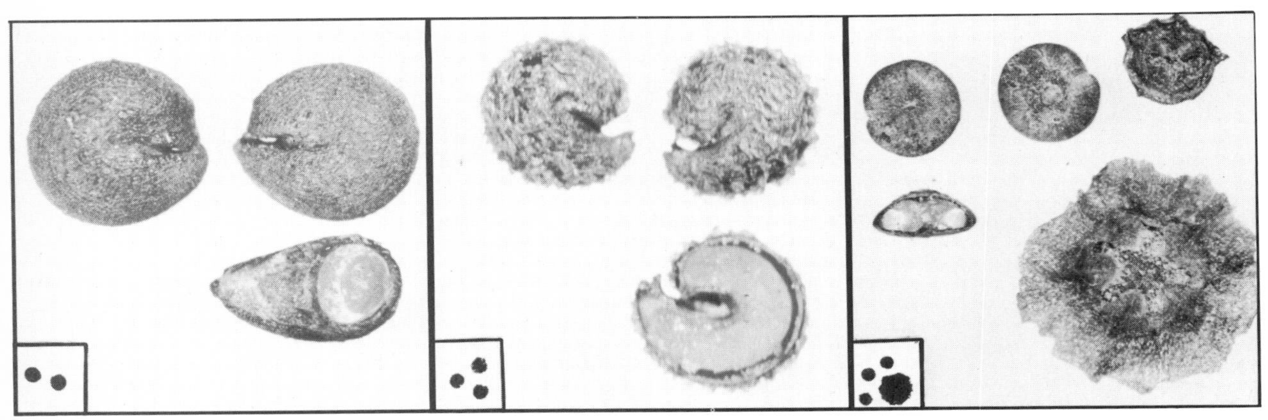

452. *Polanisia graveolens*
Clammyweed

453. *Cleome spinosa*
Spiderflower

454. *Cycloloma atriplicifolium*
Ringwing

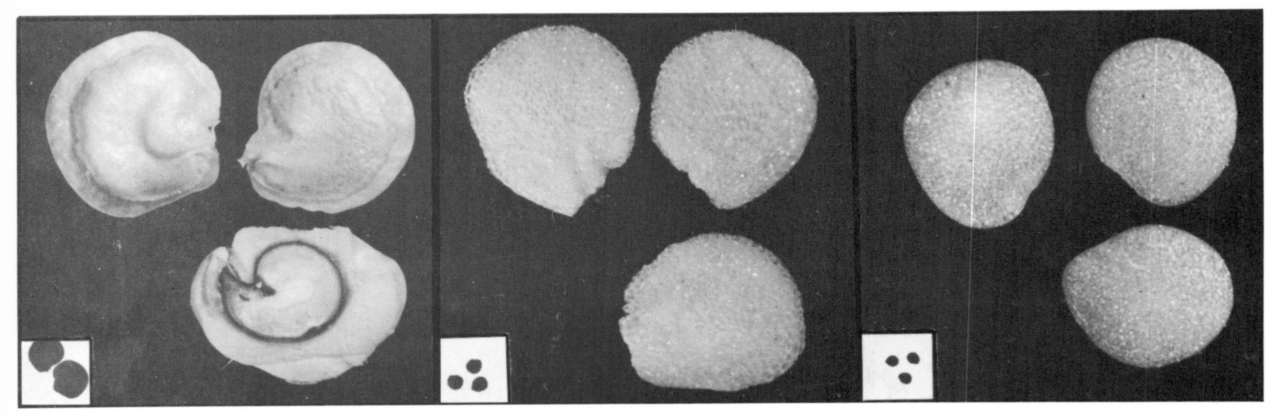

455. *Opuntia polycantha*
Pricklypear

456. *Solanum dulcamara*
Nightshade

457. *Physalis heterophylla*
Groundcherry

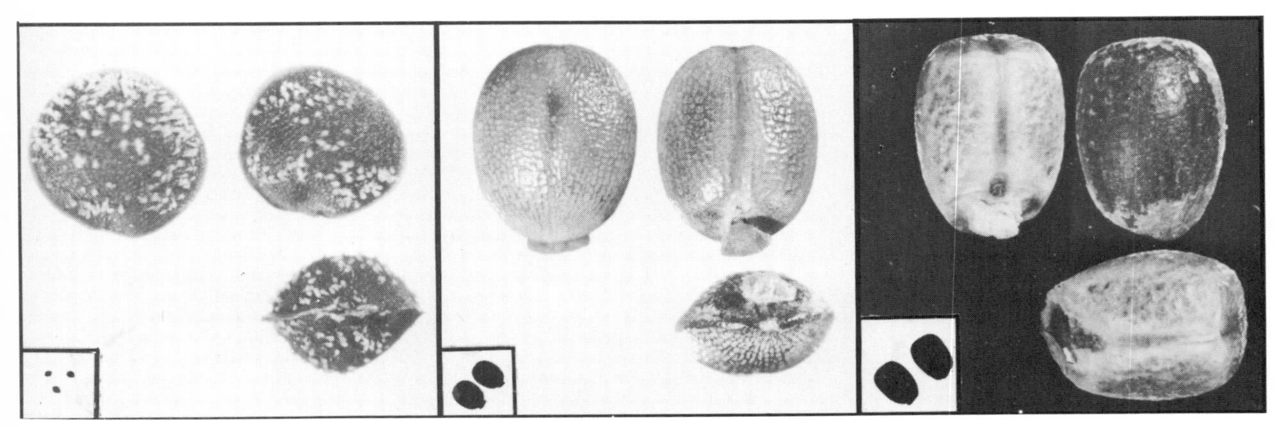

458. *Spergula arvensis*
Spurry

459. *Croton capitatus*
Doveweed

460. *Stillingia sylvatica*
Queensdelight

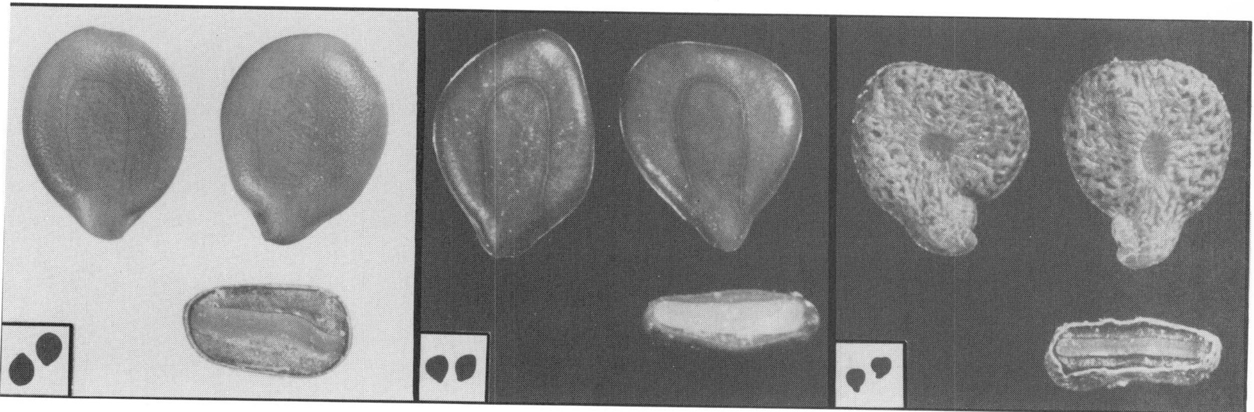

461. *Cassia occidentalis*
Senna

462. *Desmanthus illinoensis*
Bundleflower

463. *Cassia bauhinioides*
Senna

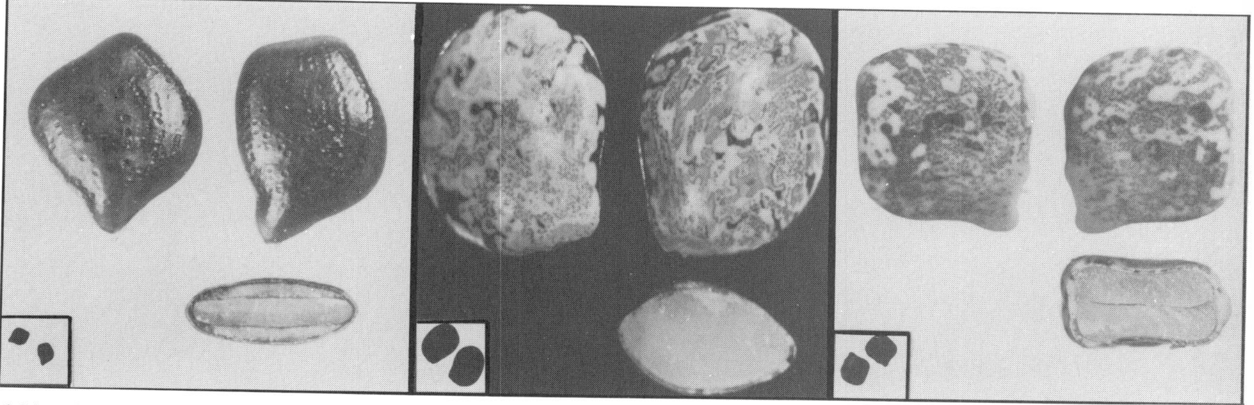

464. *Chamaecrista nictitans*
Partridgepea

465. *Lupinus albifrons*
Lupine

466. *Lupinus texensis*
Lupine

467. *Tephrosia spicata*
Hoarypea

468. *Lotus humistratus*
Deervetch

469. *Barbarea verna*
Wintercress

470. *Rhynchosia erecta*
Rhynchosia

471. *Psoralea canescens*
Scurfpea

472. *Indigofera caroliniana*
Indigo

473. *Lespedeza bicolor*
Lespedeza

474. *Galactia regularis*
Milkpea

475. *Centrosema virginianum*
Butterflypea

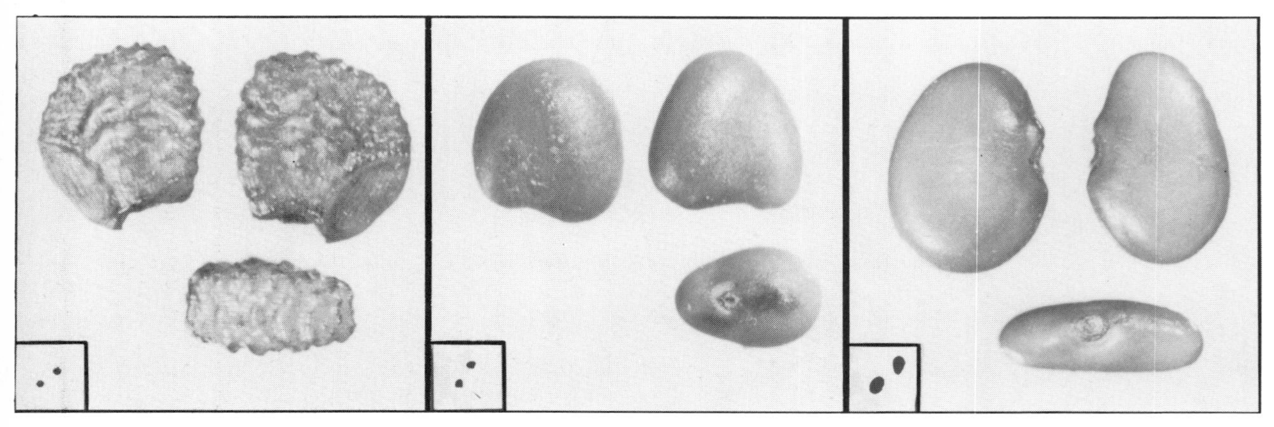

476. *Dactyloctenium aegyptium*
Crowfootgrass

477. *Trifolium hybridum*
Alsike

478. *Desmodium purpureum*
Tickclover

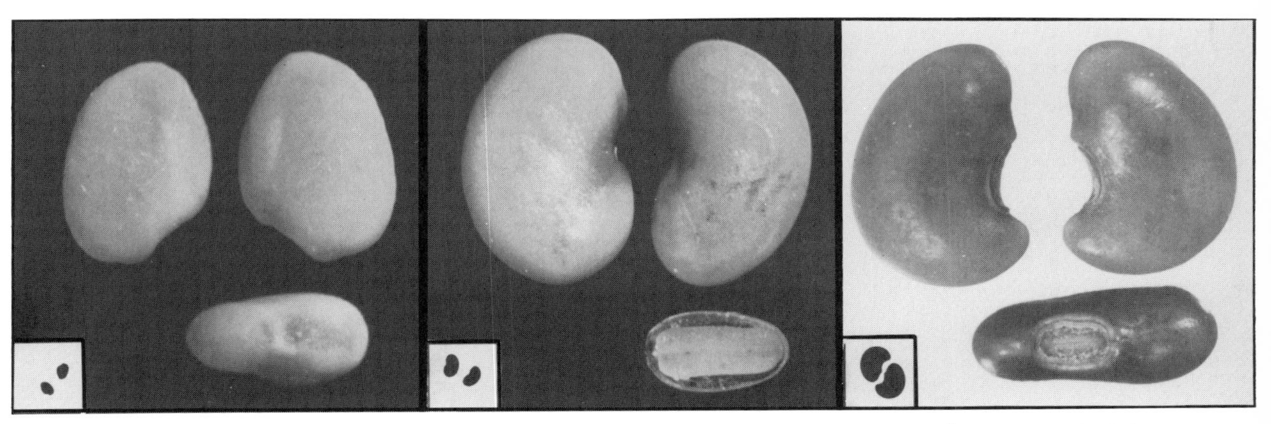

479. *Melilotus alba*
Sweetclover

480. *Medicago hispida*
Burclover

481. *Aeschynomene virginica*
Jointvetch

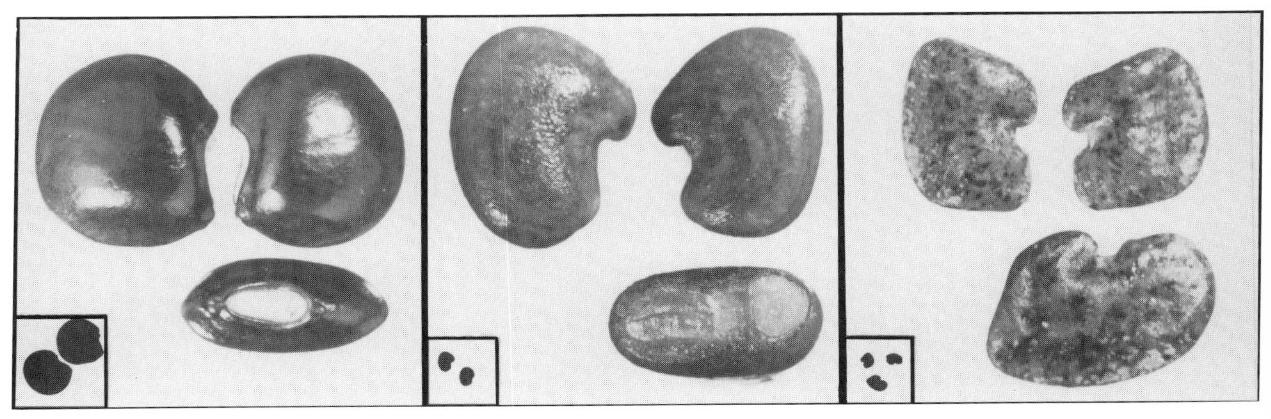

482. *Phaseolus polystachyus*
Bean

483. *Crotalaria striata*
Crotalaria

484. *Astragalus nuttallianus*
Loco

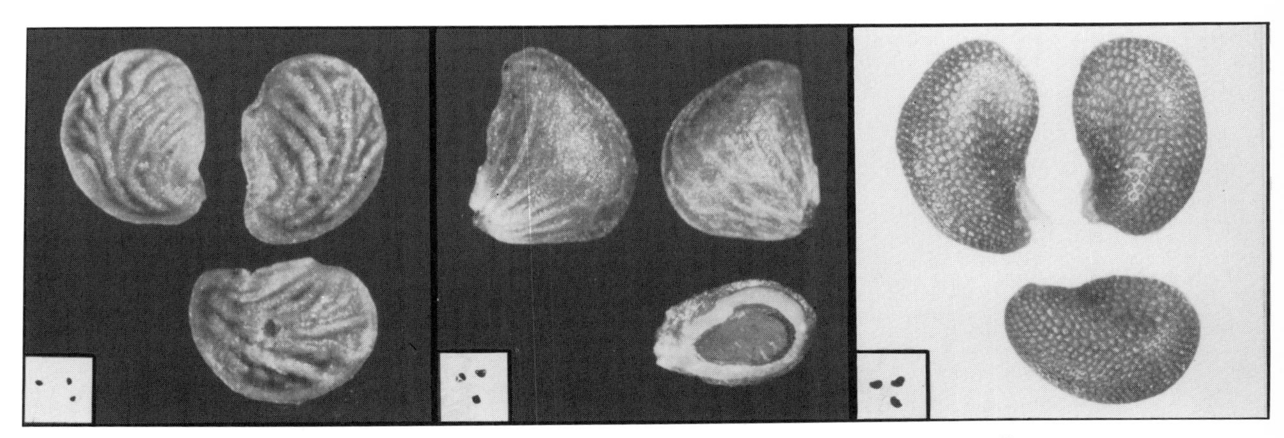

485. *Potentilla pensylvanica*
Cinquefoil

486. *Fragaria virginiana*
Strawberry

487. *Mammillaria vivipara*
Cactus

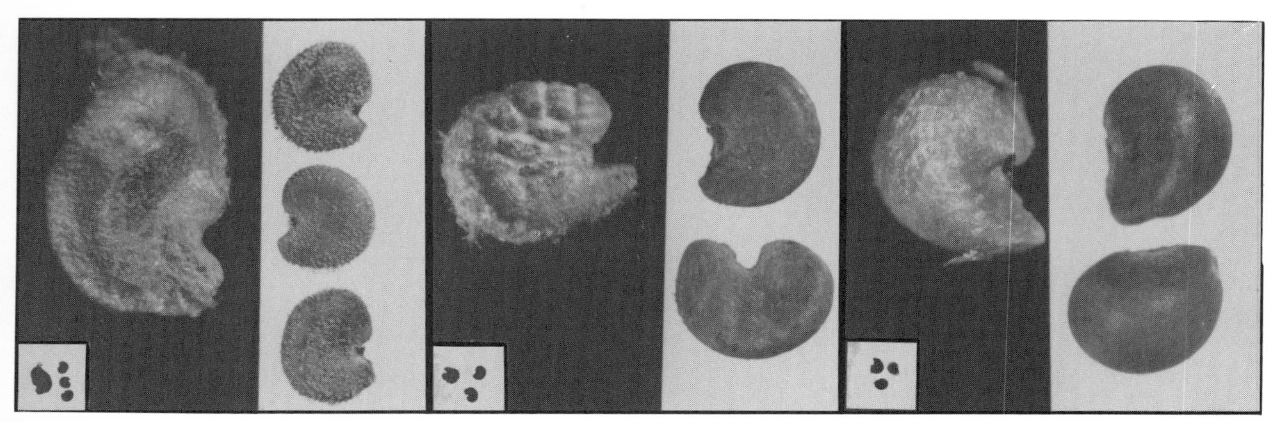

488. *Sphaeralcea fendleri*
Globemallow

489. *Malvastrum coccineum*
Falsemallow

490. *Sidalcea neomexicana*
Checkermallow

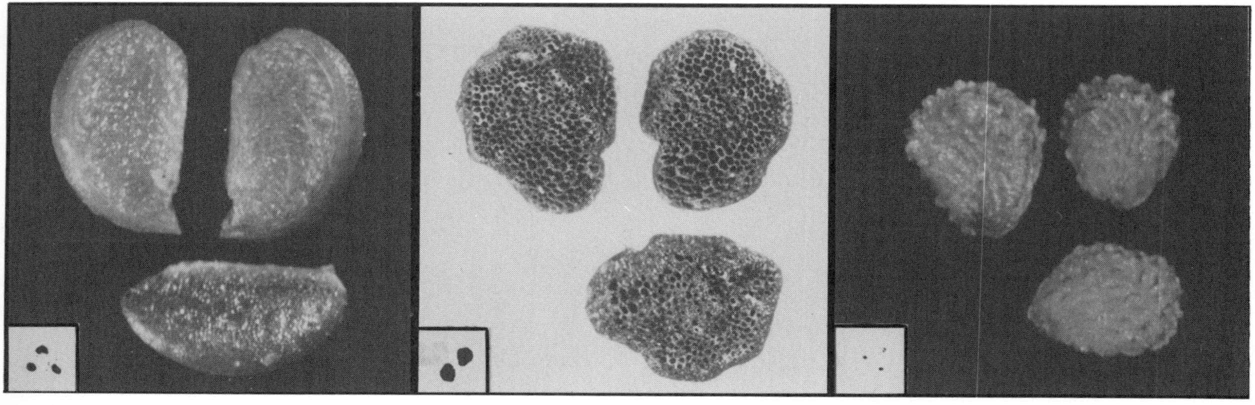

491. *Linum striatum*
Flax

492. *Solanum rostratum*
Buffalobur

493. *Cerastium viscosum*
Chickweed

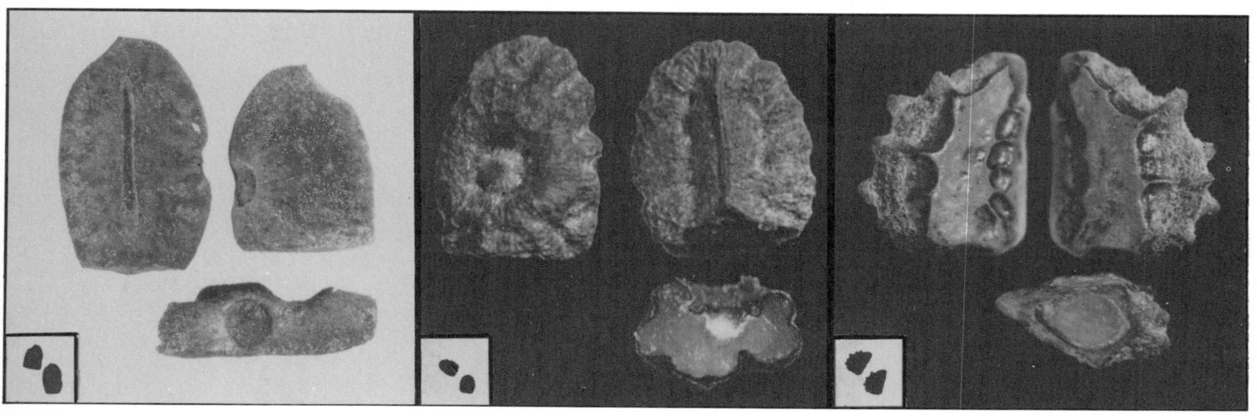

494. *Commelina virginica*
Dayflower

495. *Tradescantia virginiana*
Spiderwort

496. *Kallstroemia grandiflora*
Caltrop

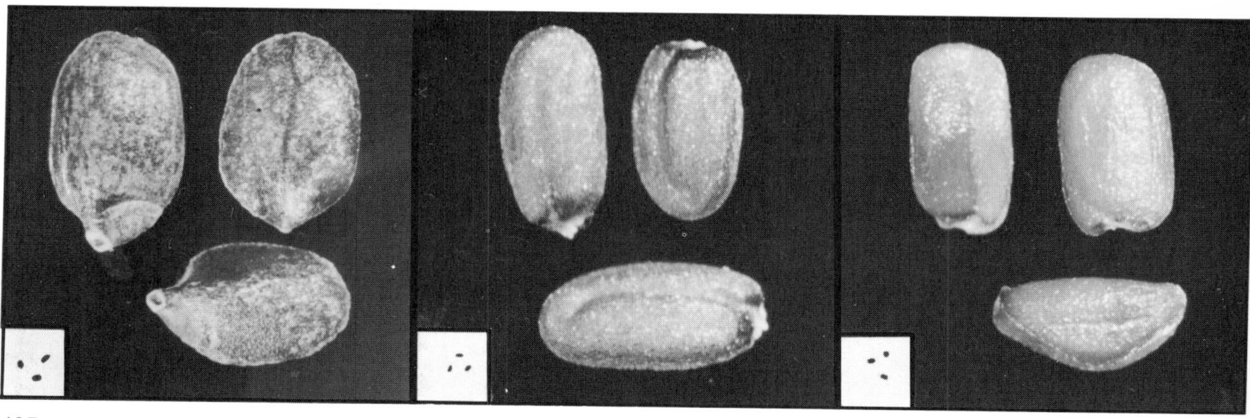

497. *Monarda punctata*
Beebalm

498. *Capsella bursa-pastoris*
Shepherds-purse

499. *Sisymbrium altissimum*
Tumblemustard

500. *Collomia grandiflora*
Collomia

501. *Phlox pilosa*
Phlox

502. *Plantago lanceolata*
Buckhorn

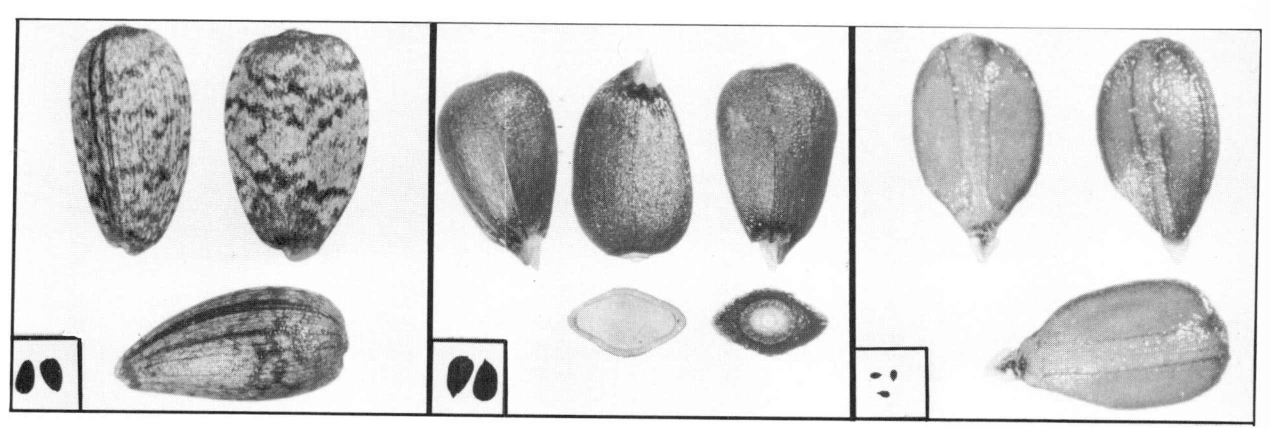

503. *Helianthus annuus*
Sunflower

504. *Centaurea americana*
Centaurea

505. *Prunella vulgaris*
Selfheal

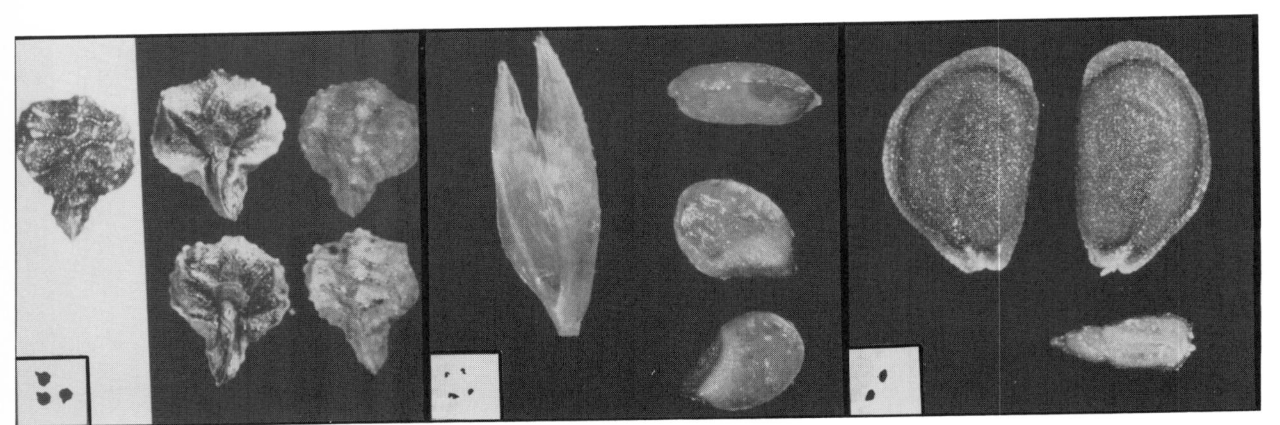

506. *Plagiobothrys nothofulvus*
Popcornflower

507. *Sporobolus cryptandrus*
Dropseed

508. *Lepidium virginicum*
Pepperweed

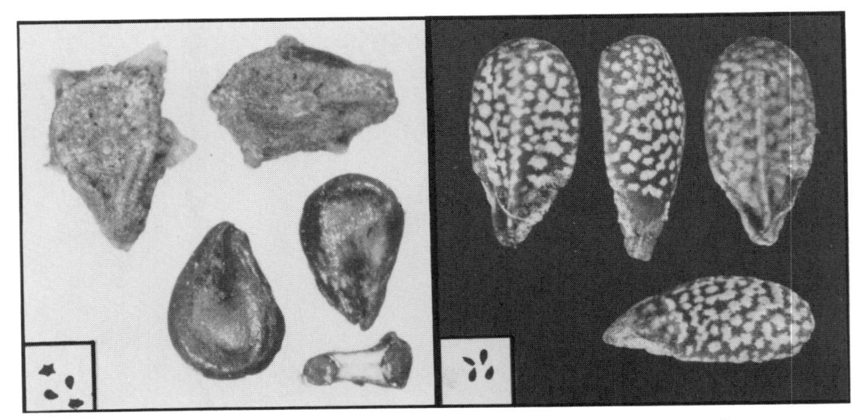

509. *Kochia scoparia*
Summer-cypress

510. *Lamium amplexicaule*
Henbit

Farmlands

Crops

Seeds of farm crops are familiar to most of us, but, because of their great importance to man and beast, a few of the principal ones are illustrated here. Corn and certain other well-known seeds which are not likely to be confused with the seeds of other species have been omitted.

Additional seed pictures of the numerous varieties of grains and other farm crops are available in many bulletins and other publications.

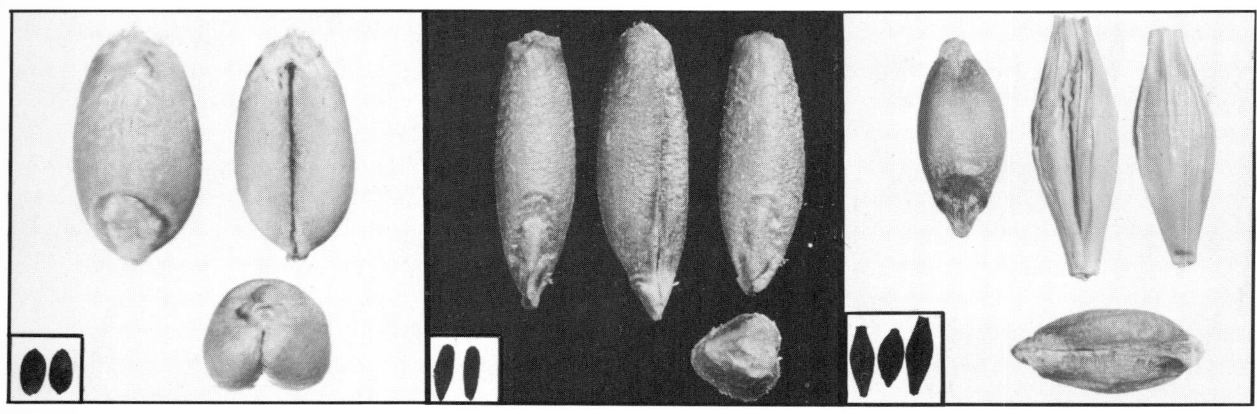

511. *Triticum aestivum*
Wheat

512. *Secale cereale*
Rye

513. *Hordeum vulgare*
Barley

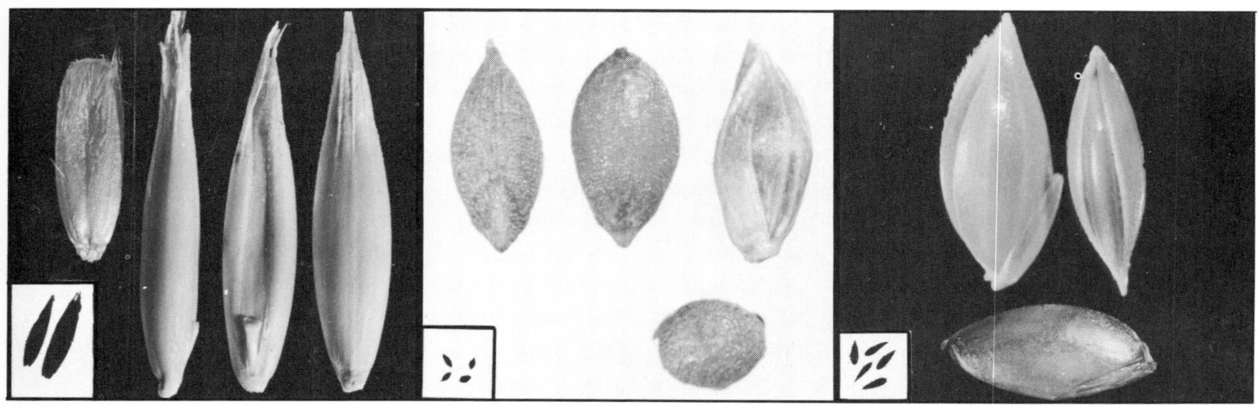

514. *Avena sativa*
Oats

515. *Phleum pratense*
Timothy

516. *Phalaris canariensis*
Canarygrass

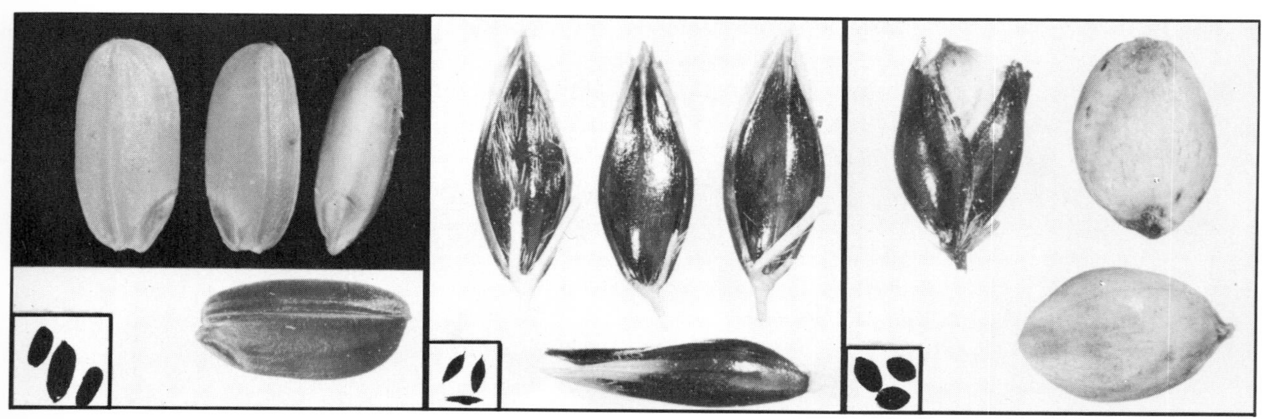

517. *Oryza sativa*
Rice

518. *Sorghum sudanense*
Sudangrass

519. *Sorghum vulgare*
Sorghum

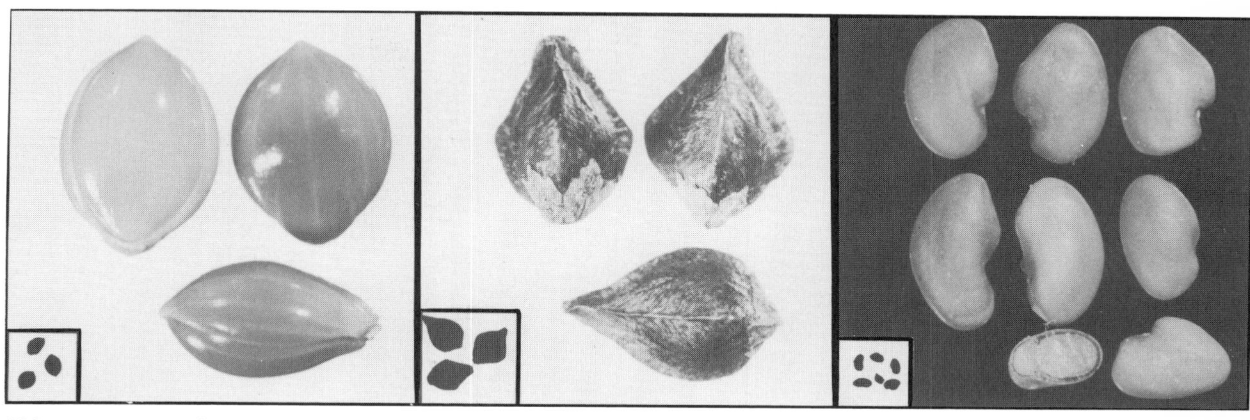

520. *Panicum miliaceum*
Proso

521. *Fagopyrum esculentum*
Buckwheat

522. *Medicago sativa*
Alfalfa

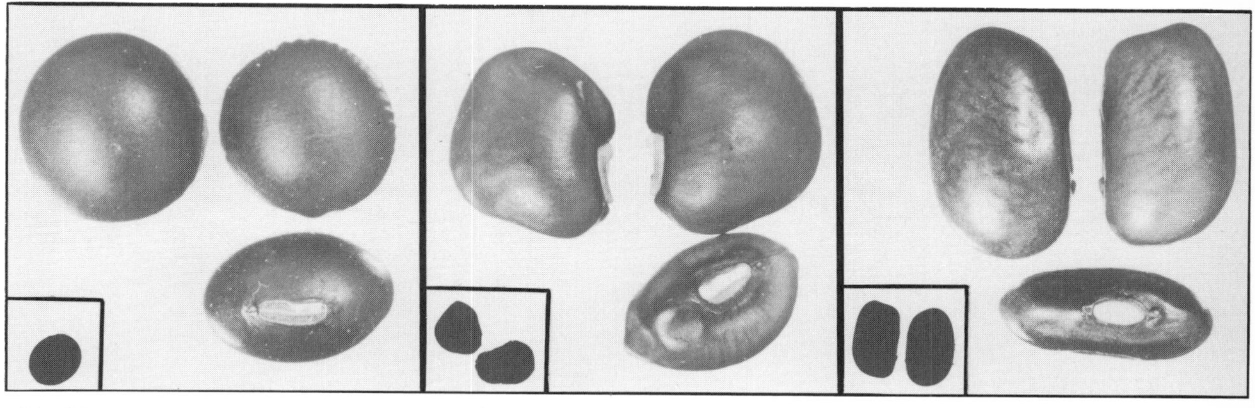

523. *Glycine max*
Soybean

524. *Vigna sinensis*
Cowpea

525. *Phaseolus vulgaris*
Bean

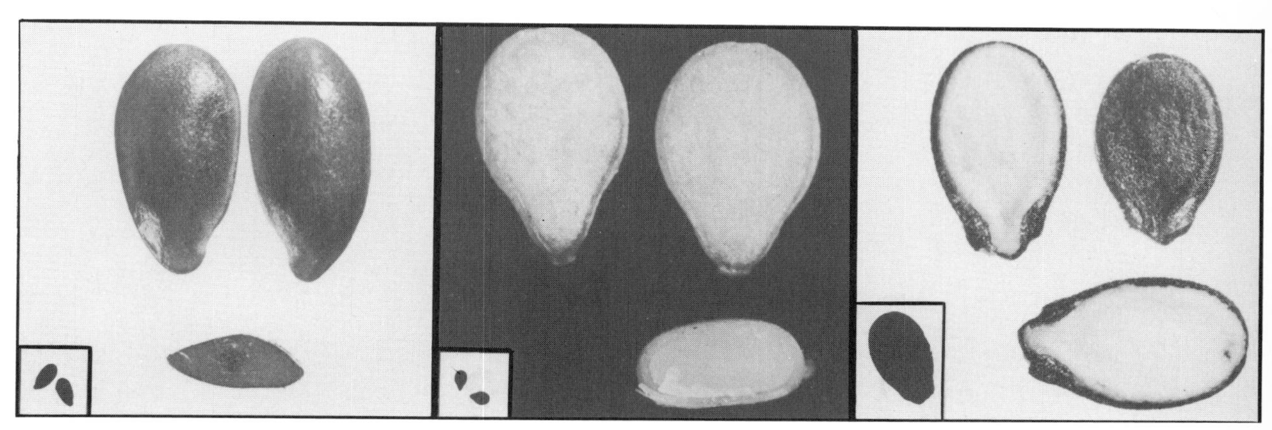

526. *Linum usitatissimum*
Flax

527. *Sesamum orientale*
Sesame

528. *Citrullus vulgaris*
Watermelon

Wetlands

Aquatics

The designation aquatics, as used here, implies plants which are submerged or have floating leaves. Of the thirty-one species illustrated the majority are useful or important as sources of food for waterfowl and other wetland wildlife.

Good illustrations of seeds of aquatics, as well as of emergent or other marsh plants, are presented in A Flora of the Marshes of California, by Herbert L. Mason, University of California Press, 1957, and also in Food of Game Ducks of the United States and Canada, by Alexander C. Martin and F. M. Uhler, United States Fish and Wildlife Service Research Report 30, 1951.

529. *Potamogeton spirillus*
Pondweed

530. *Potamogeton pusillus*
Pondweed

531. *Potamogeton foliosus*
Pondweed

532. *Potamogeton perfoliatus*
Pondweed

533. *Potamogeton epihydrus*
Pondweed

534. *Potamogeton pectinatus*
Sago Pondweed

535. *Potamogeton nodosus*
Pondweed

536. *Potamogeton natans*
Pondweed

537. *Potamogeton zosteriformis*
Pondweed

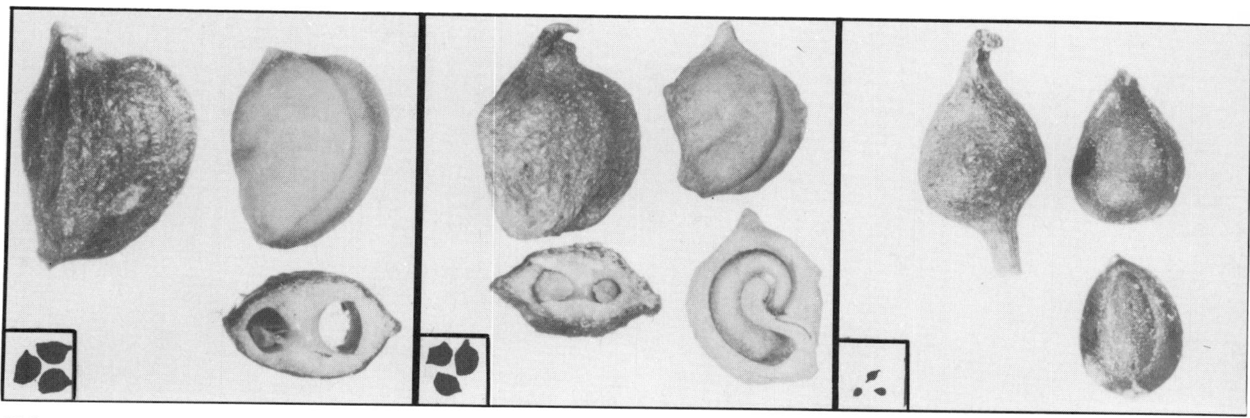

538. *Potamogeton praelongus*
Pondweed

539. *Potamogeton amplifolius*
Pondweed

540. *Ruppia maritima*
Widgeongrass

541. *Zannichellia palustris*
Hornpondweed

542. *Zostera marina*
Eelgrass

543. *Anacharis canadensis*
Waterweed

544. *Najas guadalupensis*
Naiad

545. *Najas flexilis*
Naiad

546. *Najas marina*
Naiad

547. *Vallisneria spiralis*
Wildcelery

548. *Eichhornia crassipes*
Waterhyacinth

549. *Ceratophyllum demersum*
Coontail

550. *Nuphar luteum*
Spatterdock

551. *Nymphaea odorata*
Waterlily

552. *Nymphaea tuberosa*
Waterlily

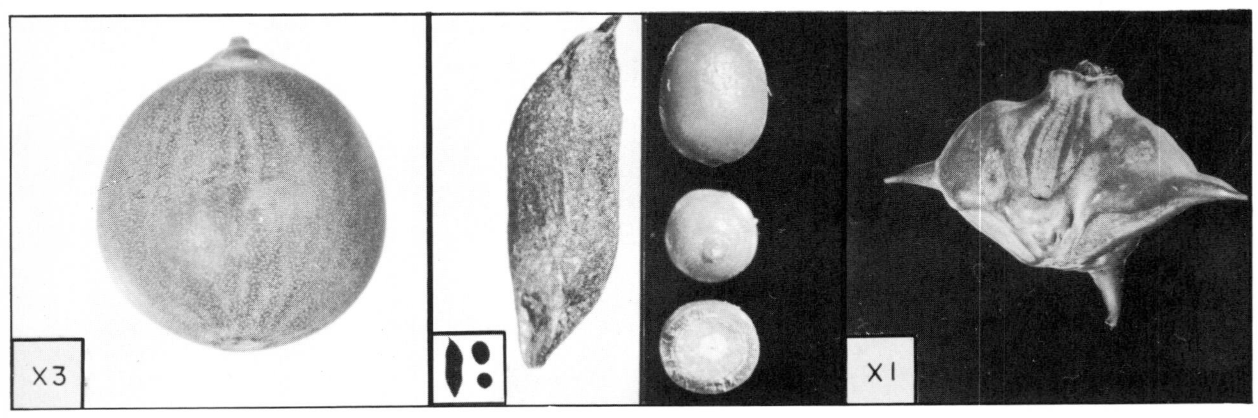

553. *Nelumbo lutea*
Lotus

554. *Brasenia schreberi*
Watershield

555. *Trapa natans*
Waterchestnut

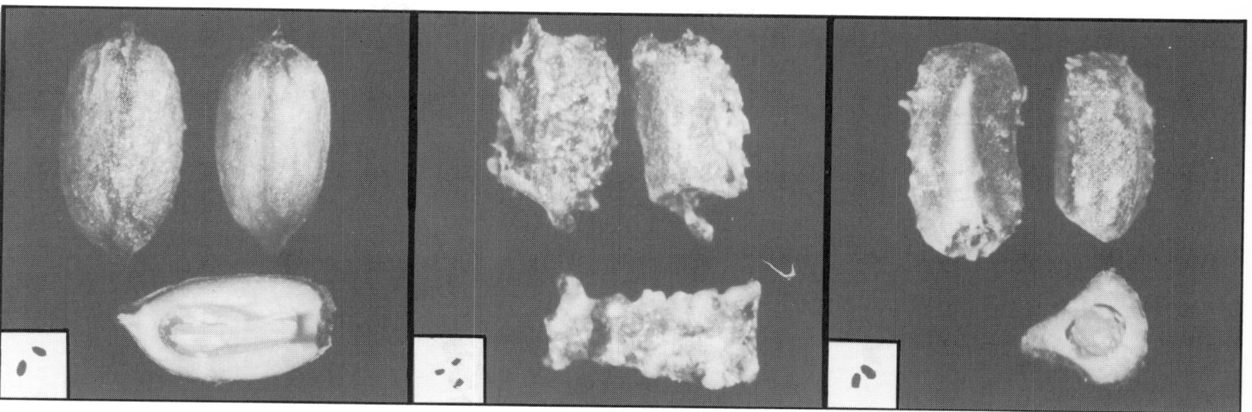

556. *Hippuris vulgaris*
Marestail

557. *Myriophyllum scabratum*
Watermilfoil

558. *Myriophyllum spicatum*
Watermilfoil

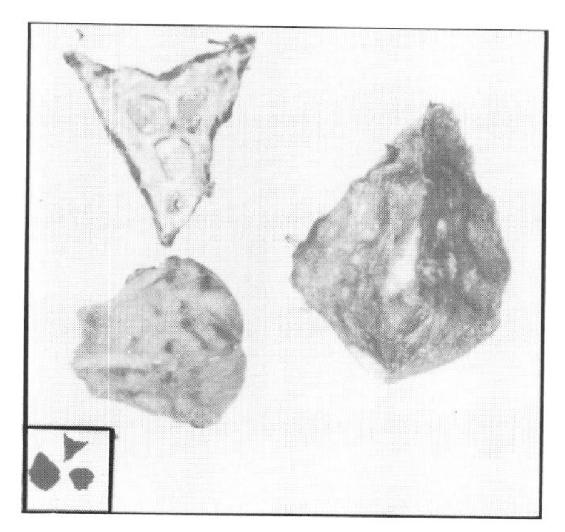

559. *Proserpinaca palustris*
Mermaidweed

Wetlands

Marsh Grasses

The eighteen species of marsh grasses illustrated here are among the most abundant and widespread marsh inhabitants of the grass family in the United States. They have varying degrees of value for waterfowl and other wildlife. A few kinds that have only limited usefulness are all too plentiful in some localities and deserve to be regarded as pest plants in such places.

560. *Glyceria striata*
Mannagrass

561. *Glyceria grandis*
Mannagrass

562. *Leptochloa fascicularis*
Sprangletop

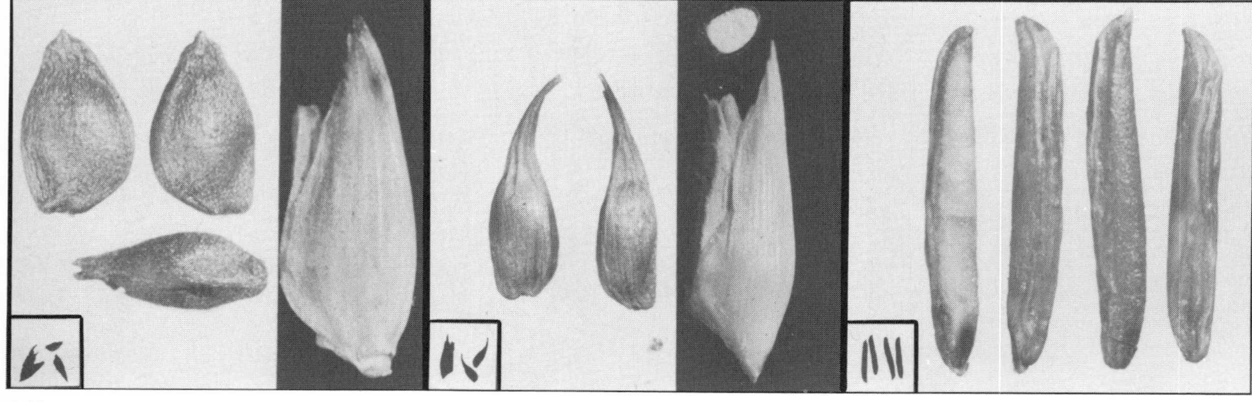

563. *Distichlis spicata*
Saltgrass

564. *Distichlis stricta*
Saltgrass

565. *Spartina alterniflora*
Cordgrass

566. *Phalaris arundinacea*
Canarygrass

567. *Leersia oryzoides*
Rice Cutgrass

568. *Zizania aquatica*
Wildrice

569. *Zizaniopsis miliacea*
Giant Cutgrass

570. *Paspalum distichum*
Knotgrass

571. *Panicum agrostoides*
Panicum

572. *Panicum verrucosum*
Wartgrass

573. *Panicum virgatum*
Switchgrass

574. *Panicum fasciculatum*
Browntop

575. *Echinochloa colonum*
Wildmillet

576. *Echinochloa walteri*
Wildmillet

577. *Echinochloa crusgalli*
Wildmillet

Wetlands

Marsh Sedges

The name sedge is often used for the genus *Carex,* but for the purposes of this manual its implications are broadened to apply to the Cyperaceae as a whole, as the sedge family. Many of the genera and species of this large family grow in marshes, and most of the forty-two species illustrated are useful as food or cover for waterfowl and other wetland wildlife.

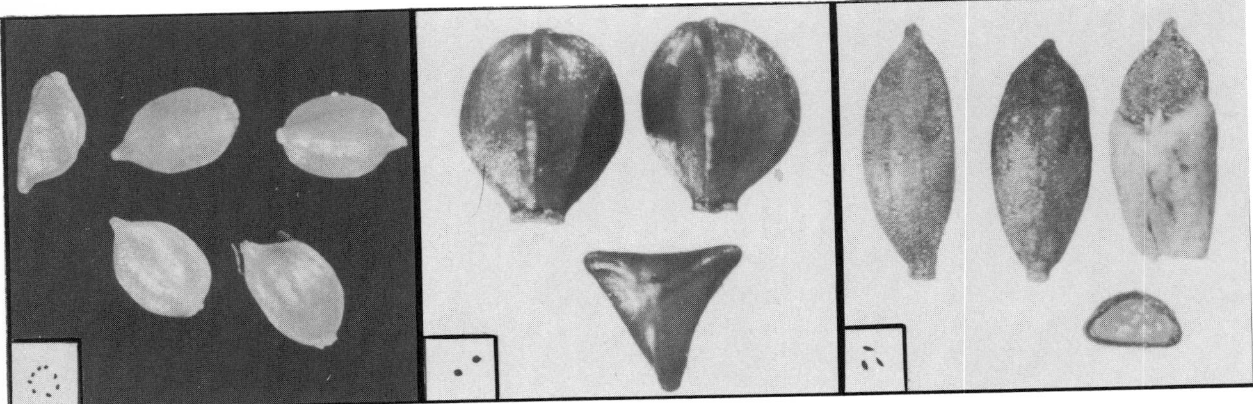

578. *Cyperus erythrorhizos*
Flatsedge

579. *Cyperus compressus*
Flatsedge

580. *Cyperus odoratus*
Flatsedge

581. *Cyperus strigosus*
Flatsedge

582. *Cyperus esculentus*
Chufa

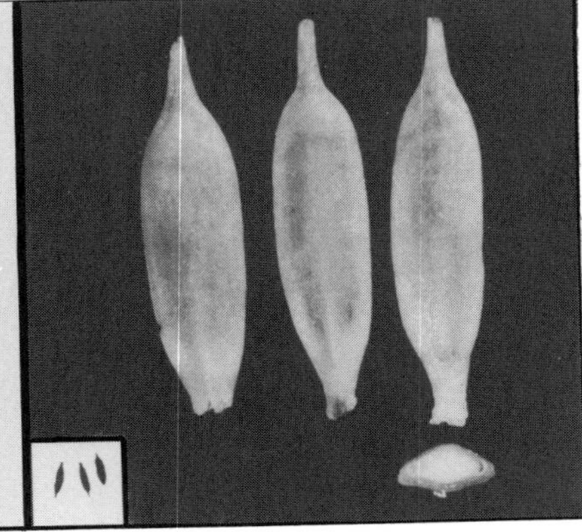

583. *Dulichium arundinaceum*
Dulichium

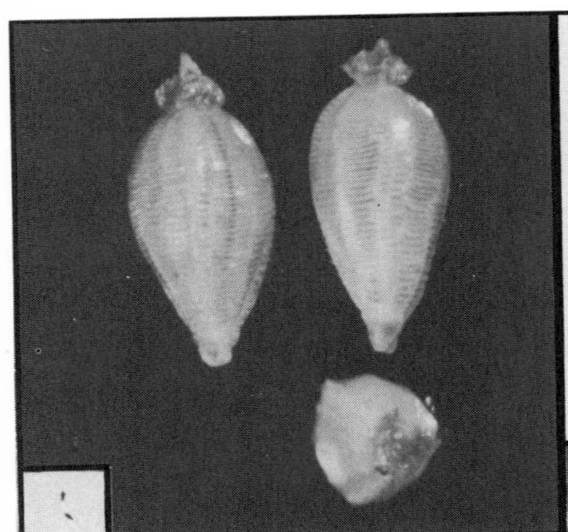

584. *Eleocharis acicularis*
Spikerush

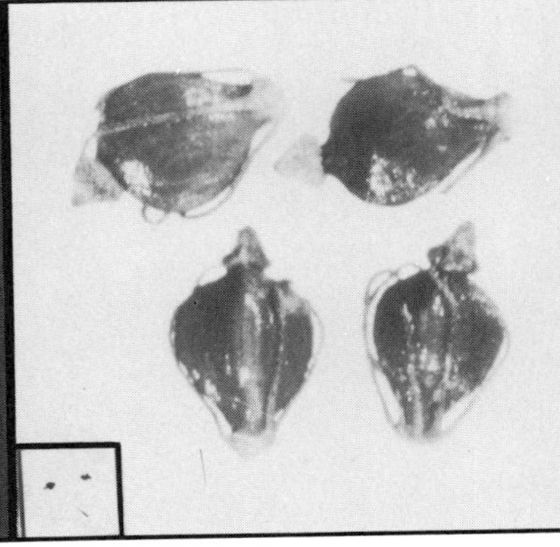

585. *Eleocharis albida*
Spikerush

586. *Eleocharis olivacea*
Spikerush

587. *Eleocharis ovata*
Spikerush

588. *Eleocharis obtusa*
Spikerush

589. *Eleocharis parvula*
Spikerush

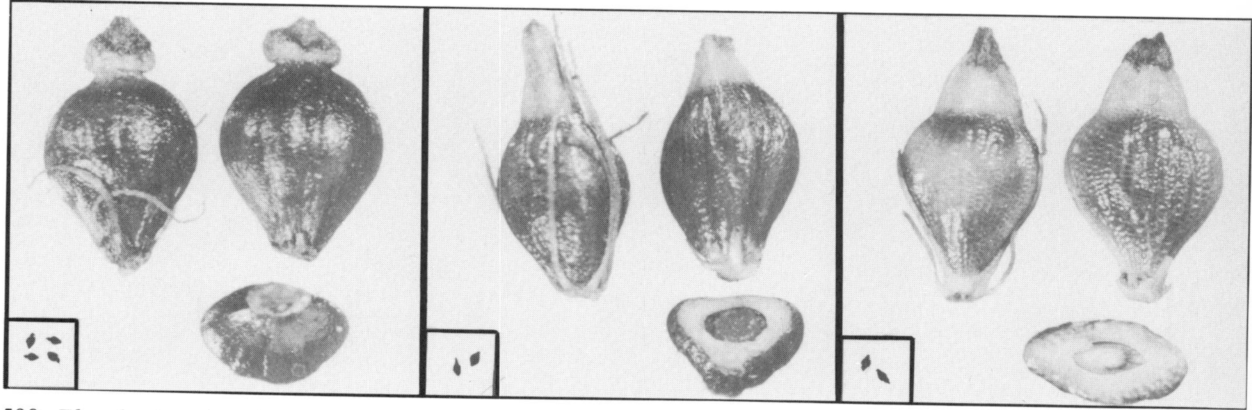

590. *Eleocharis palustris*
Spikerush

591. *Eleocharis rostellata*
Spikerush

592. *Eleocharis cellulosa*
Spikerush

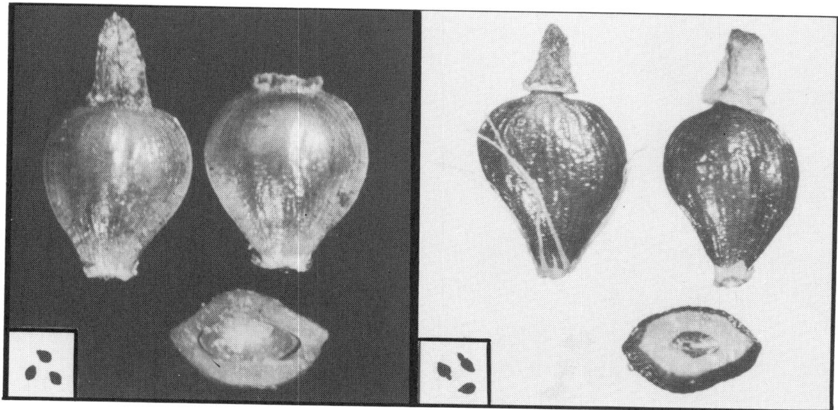

593. *Eleocharis equisetoides*
Spikerush

594. *Eleocharis quadrangulata*
Spikerush

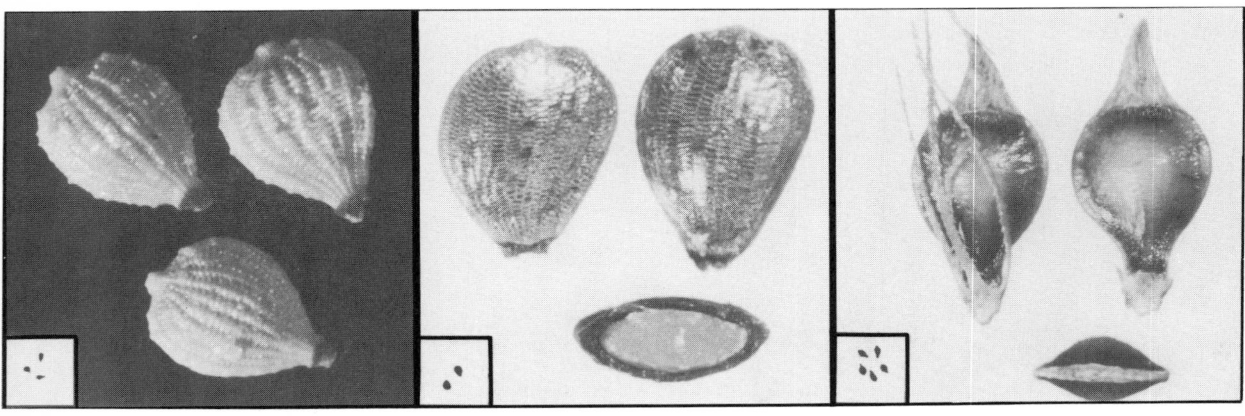

595. *Fimbristylis baldwiniana*
Fimbristylis

596. *Fimbristylis castanea*
Fimbristylis

597. *Rhynchospora capitellata*
Beakrush

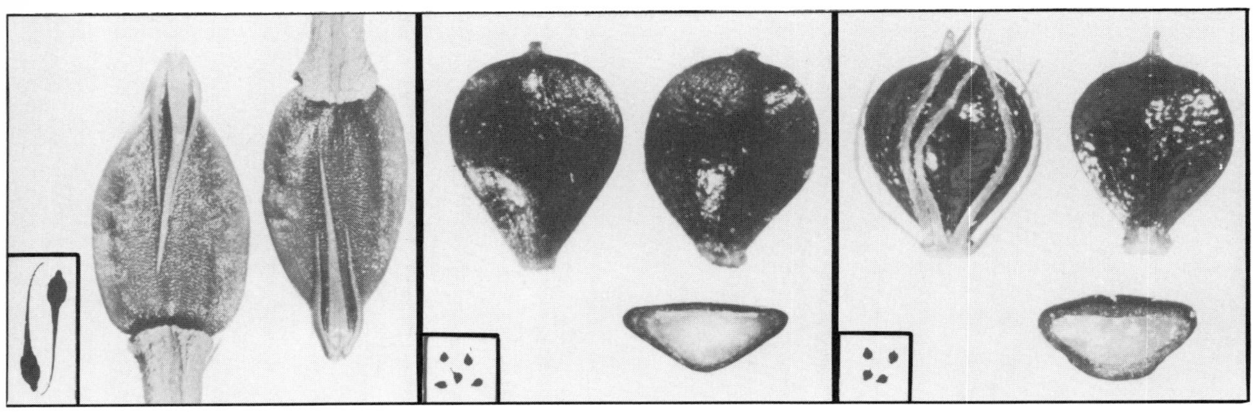

598. *Rhynchospora corniculata*
Beakrush

599. *Scirpus smithii*
Bulrush

600. *Scirpus debilis*
Bulrush

601. *Scirpus californicus*
Bulrush

602. *Scirpus validus*
Bulrush

603. *Scirpus acutus*
Bulrush

MARSH SEDGES

PLATES 604–612

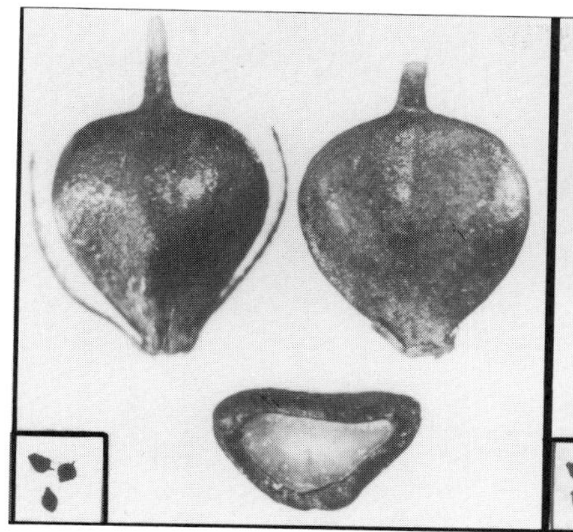

604. *Scirpus heterochaetus*
Bulrush

605. *Scirpus subterminalis*
Bulrush

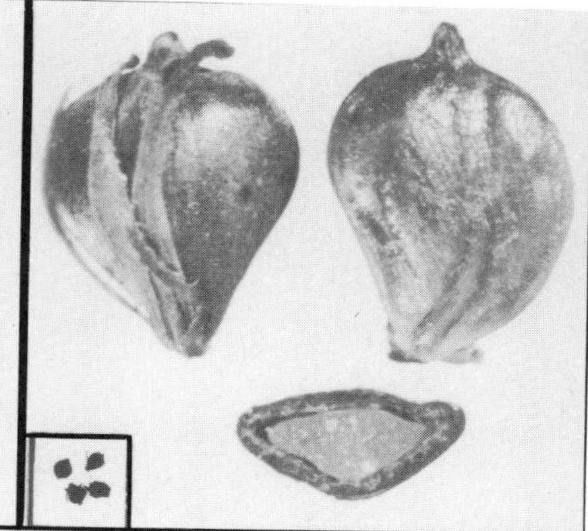

606. *Scirpus olneyi*
Bulrush

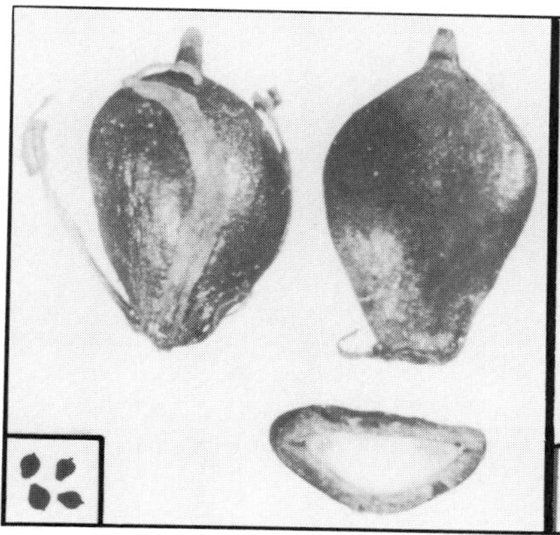

607. *Scirpus americanus*
Bulrush

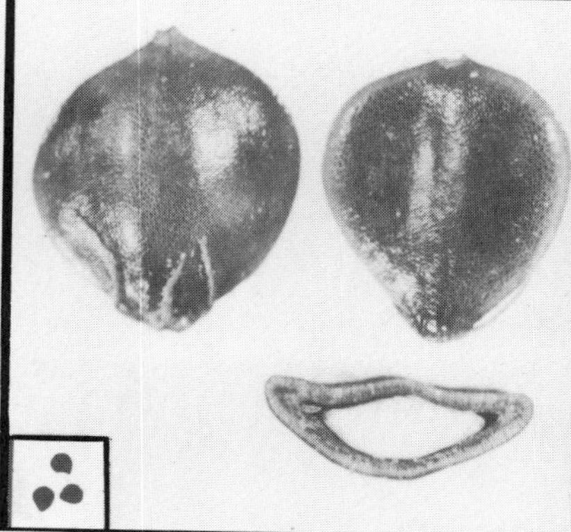

608. *Scirpus nevadensis*
Bulrush

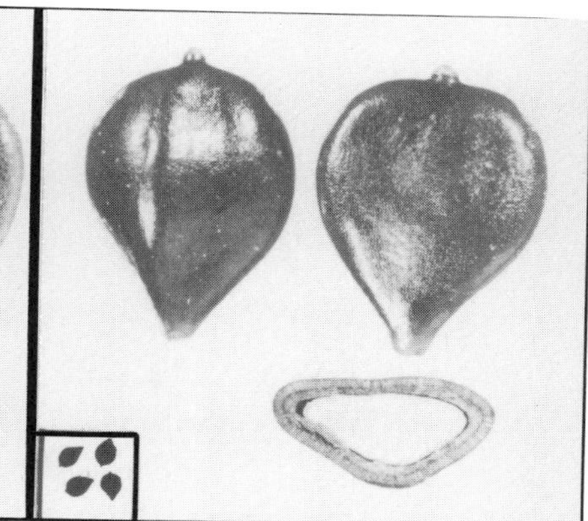

609. *Scirpus robustus*
Bulrush

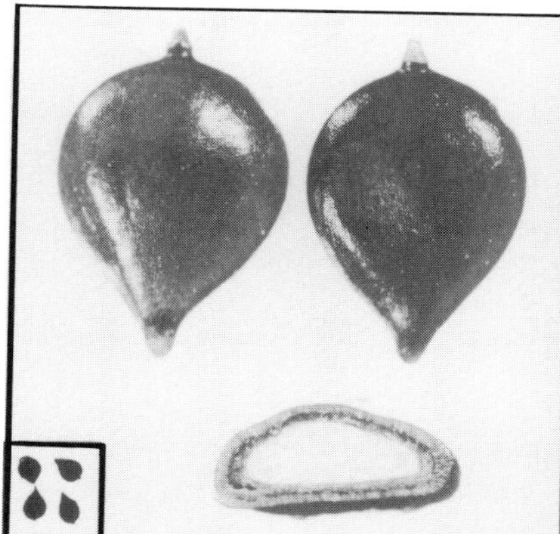

610. *Scirpus paludosus*
Bulrush

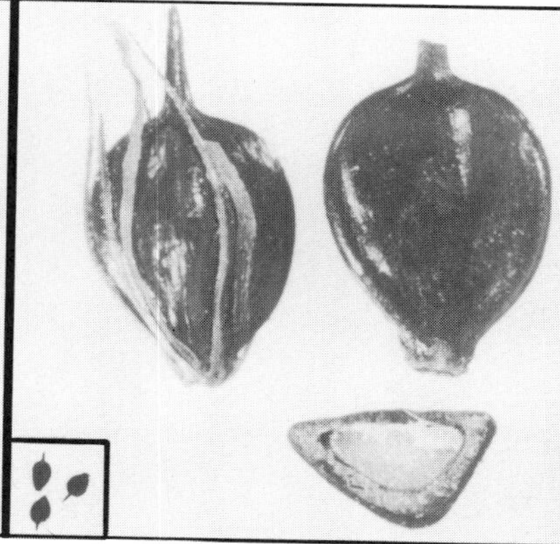

611. *Scirpus etuberculatus*
Bulrush

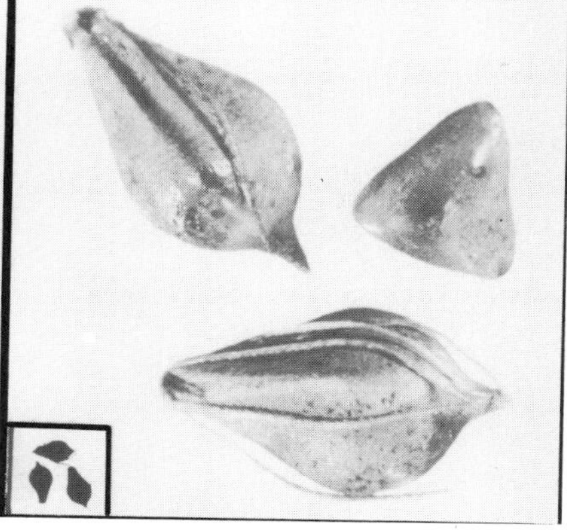

612. *Scirpus fluviatilis*
Bulrush

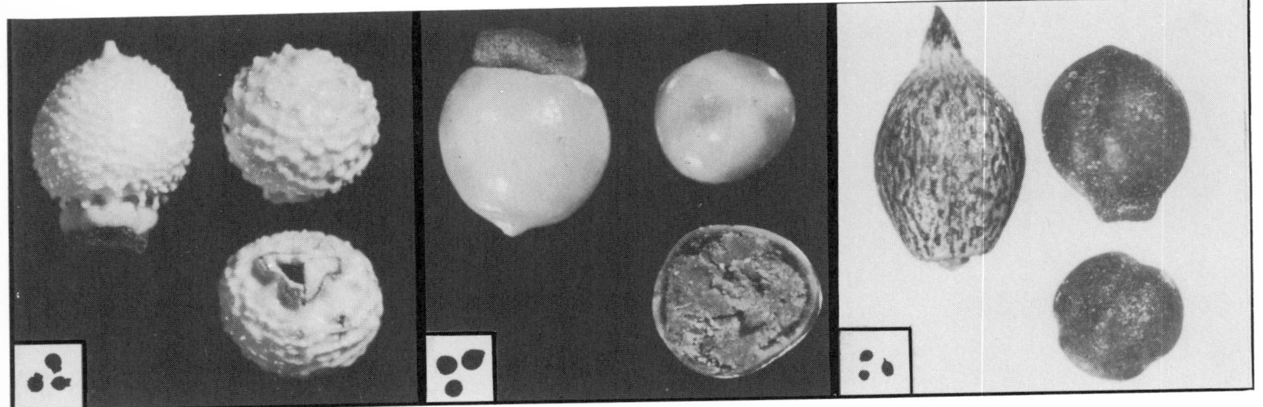

613. *Scleria ciliata*
Scleria

614. *Scleria triglomerata*
Scleria

615. *Cladium jamaicense*
Sawgrass

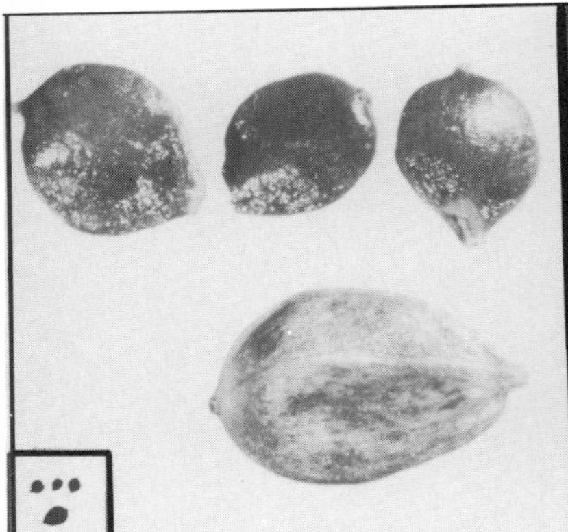

616. *Carex aquatilis*
Sedge

617. *Carex comosa*
Sedge

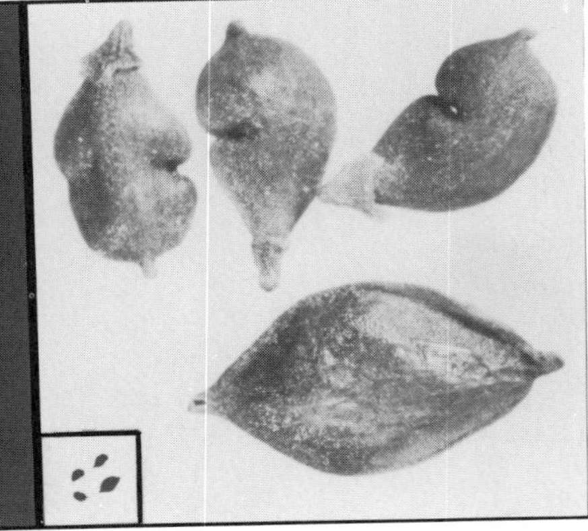

618. *Carex crinita*
Sedge

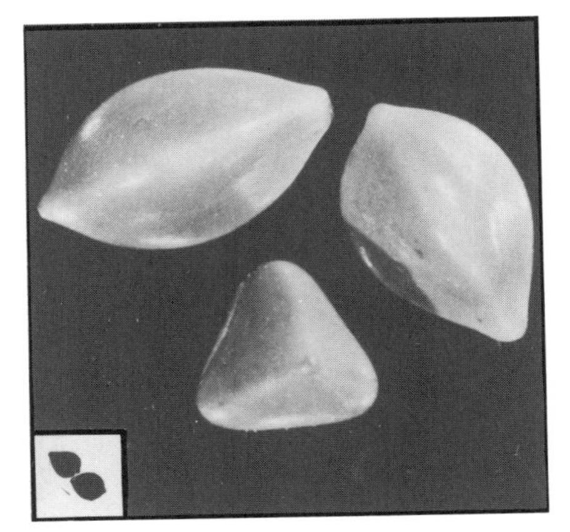

619. *Carex intumescens*
Sedge

Wetlands

Other Marsh Plants

This rather miscellaneous category is made up of the fifty-two illustrated marshland or wetland plants which do not properly belong under the headings of Aquatics, Marsh Grasses, or Marsh Sedges as those terms are defined above. Most of the plants in this section are significant to wildlife for food or protective cover, or both. Of the smartweeds (*Polygonum*) pictured, some grow in seasonally wet fields and therefore could also be included among farm weeds.

620. *Typha domingensis*
Cattail

621. *Sparganium chlorocarpum*
Burreed

622. *Sparganium eurycarpum*
Burreed

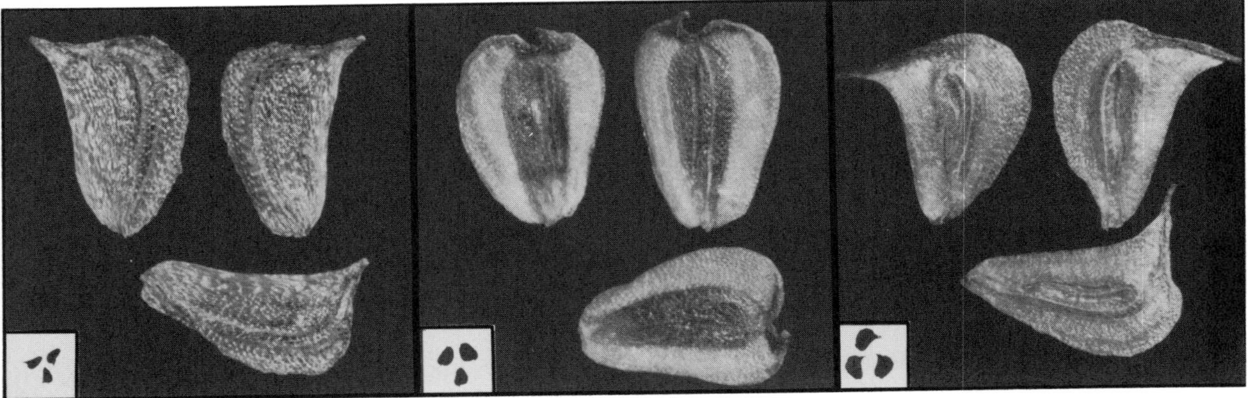

623. *Sagittaria platyphylla*
Arrowhead

624. *Sagittaria cuneata*
Arrowhead

625. *Sagittaria latifolia*
Arrowhead

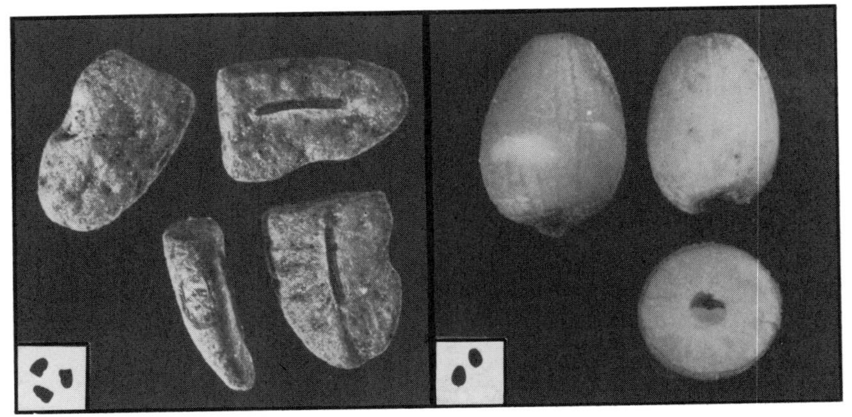

626. *Aneilema keisak*
Aneilema

627. *Pontederia cordata*
Pickerelweed

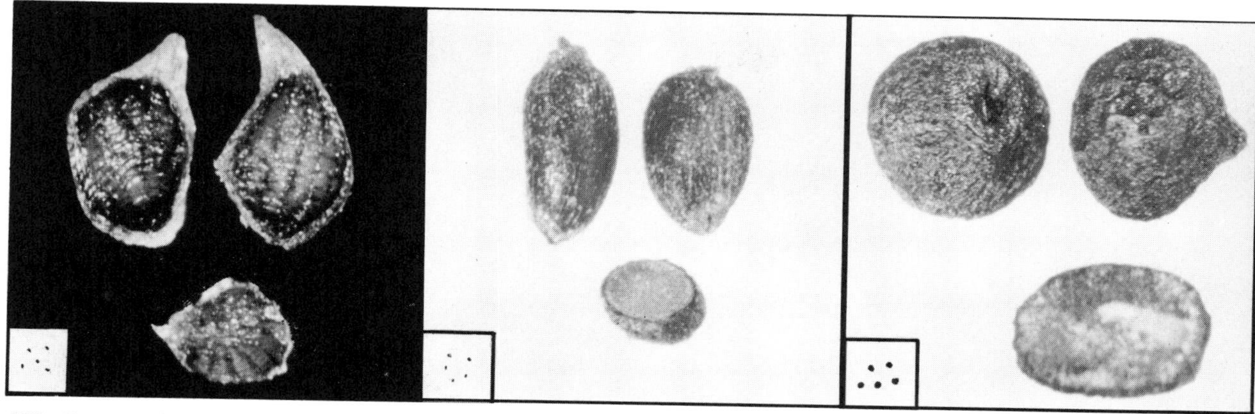

628. *Juncus effusus*
Rush

629. *Juncus roemerianus*
Needlerush

630. *Sisyrinchium graminoides*
Blue-eyed-grass

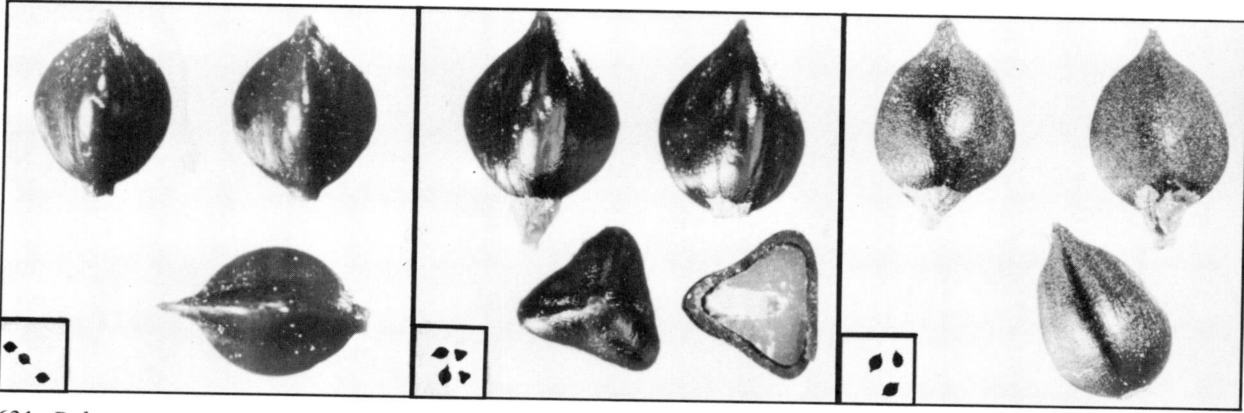

631. *Polygonum hydropiperoides*
Smartweed

632. *Polygonum punctatum*
Smartweed

633. *Polygonum hydropiper*
Smartweed

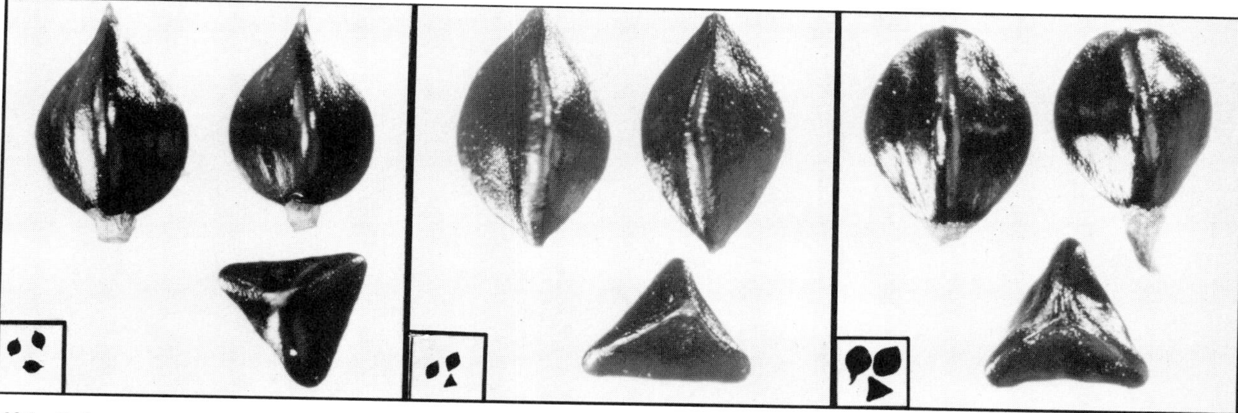

634. *Polygonum sagittatum*
Tearthumb

635. *Polygonum dumetorum*
Knotweed

636. *Polygonum scandens*
Knotweed

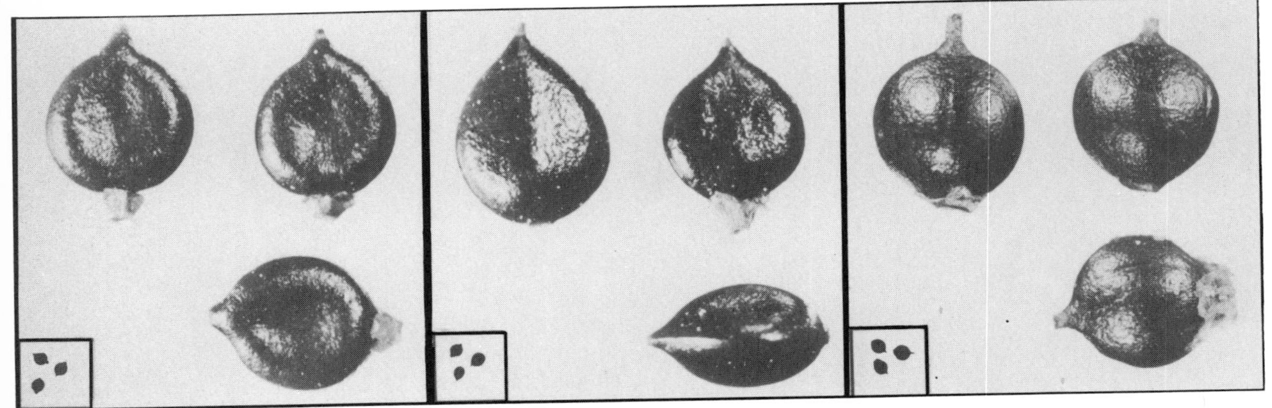

637. *Polygonum lapathifolium*
Smartweed

638. *Polygonum persicaria*
Smartweed

639. *Polygonum amphibium*
Smartweed

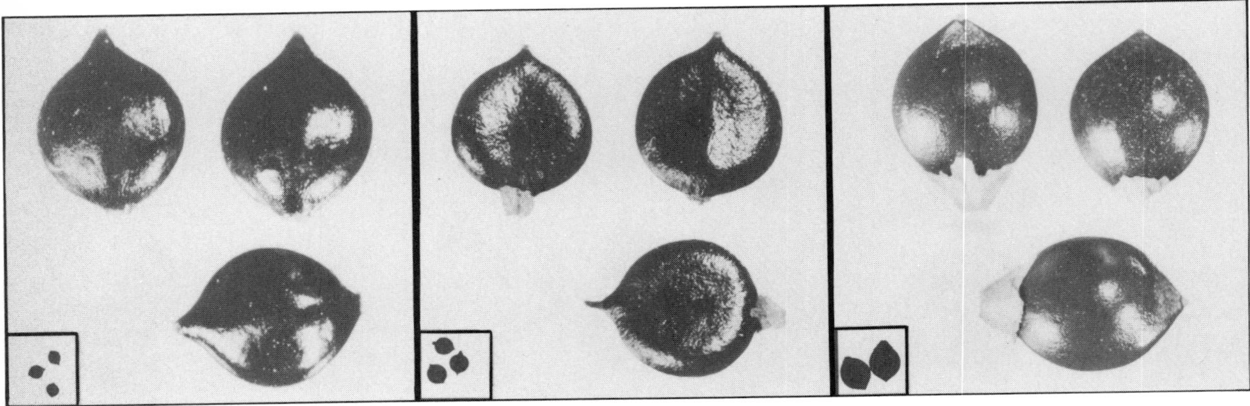

640. *Polygonum densiflorum*
Smartweed

641. *Polygonum pensylvanicum*
Smartweed

642. *Polygonum arifolium*
Tearthumb

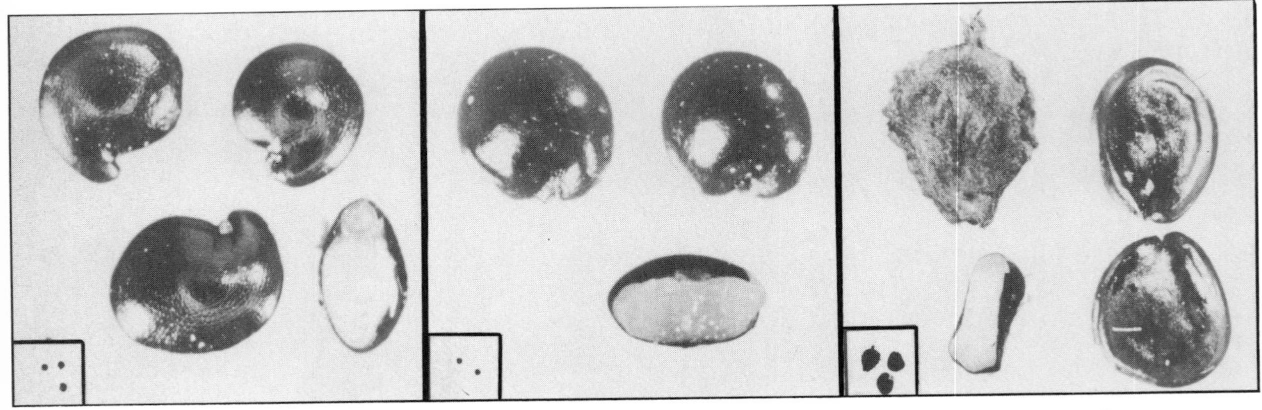

643. *Suaeda depressa*
Seablite

644. *Amaranthus tamariscinus*
Waterhemp

645. *Amaranthus cannabinus*
Waterhemp

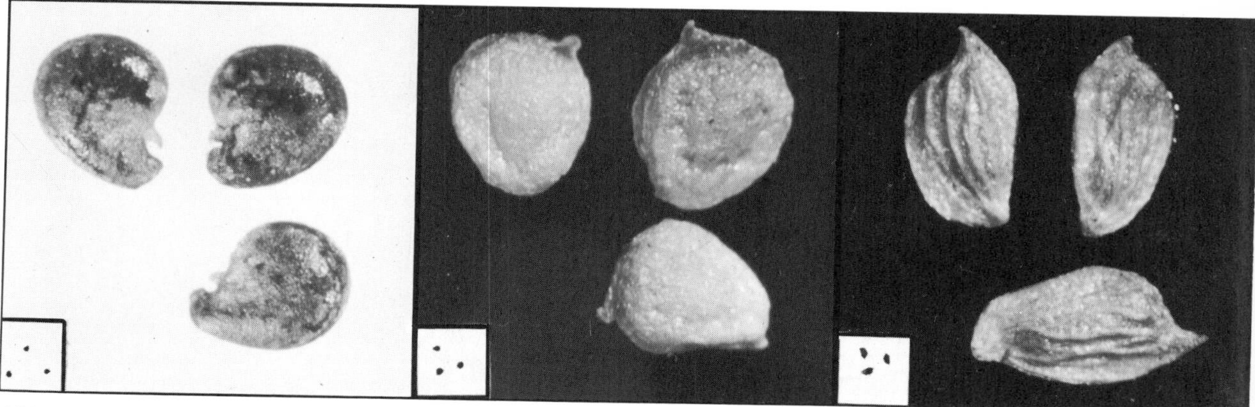

646. *Sesuvium sessile*
Sea-purslane

647. *Ranunculus sceleratus*
Buttercup

648. *Ranunculus cymbalaria*
Buttercup

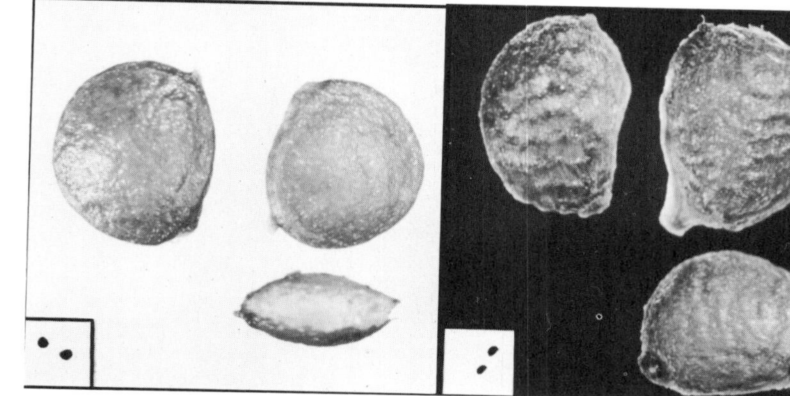

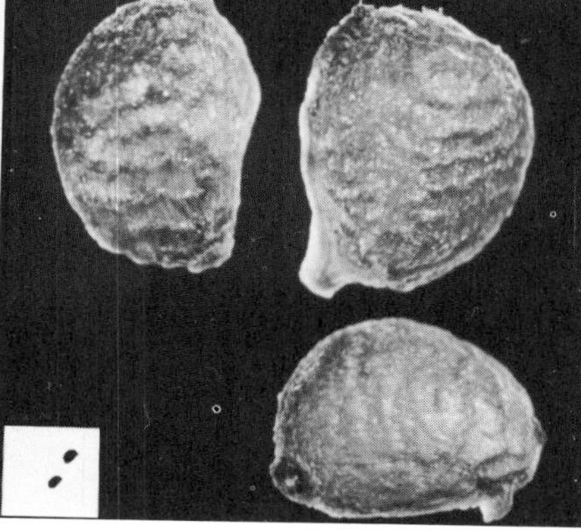

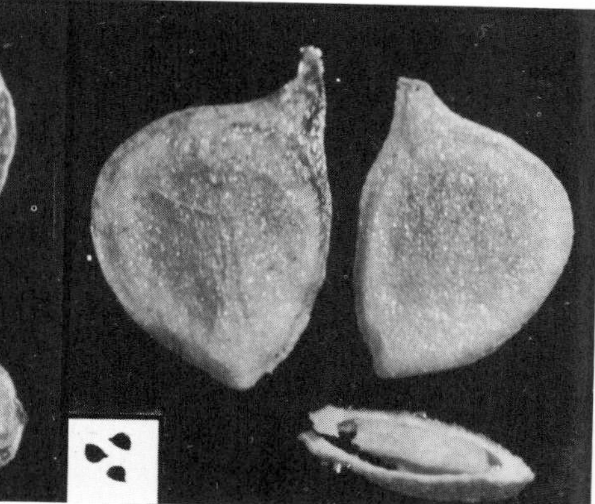

649. *Ranunculus abortivus*
Buttercup

650. *Ranunculus trichophyllus*
Buttercup

651. *Ranunculus pensylvanicus*
Buttercup

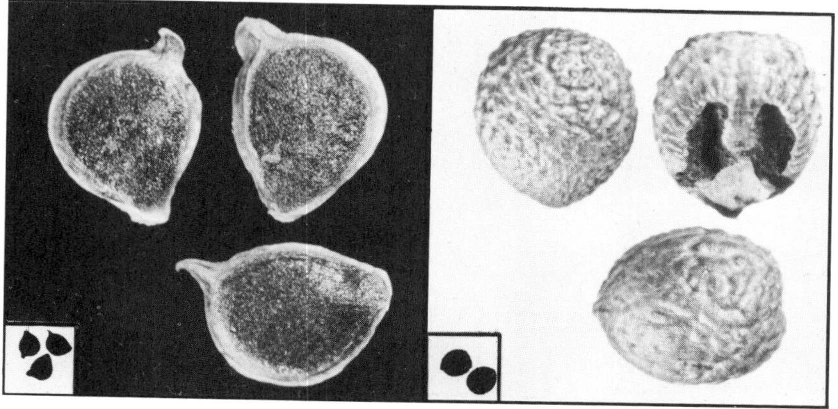

652. *Ranunculus bulbosus*
Buttercup

653. *Stillingia aquatica*
Queensdelight

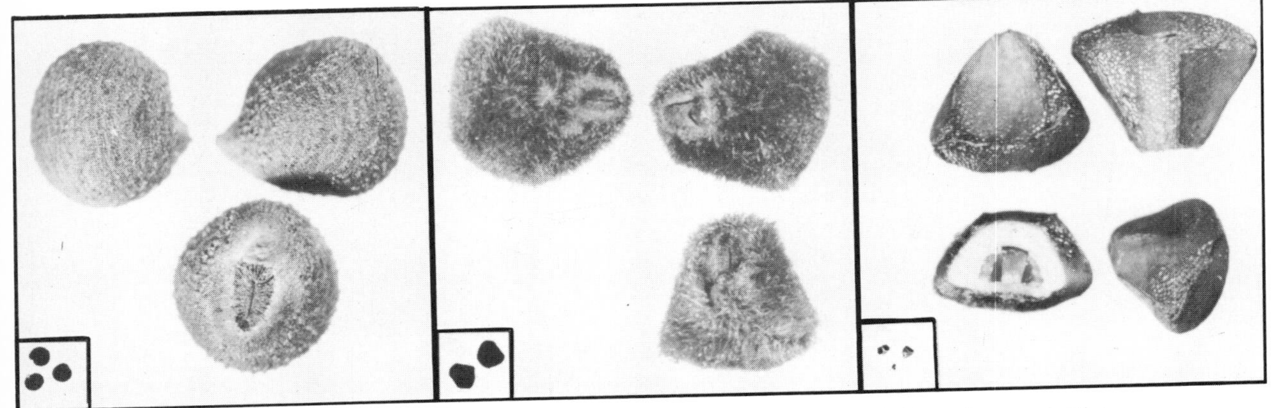

654. *Hibiscus oculiroseus*
Rosemallow

655. *Hibiscus militaris*
Rosemallow

656. *Decodon verticillatus*
Waterwillow

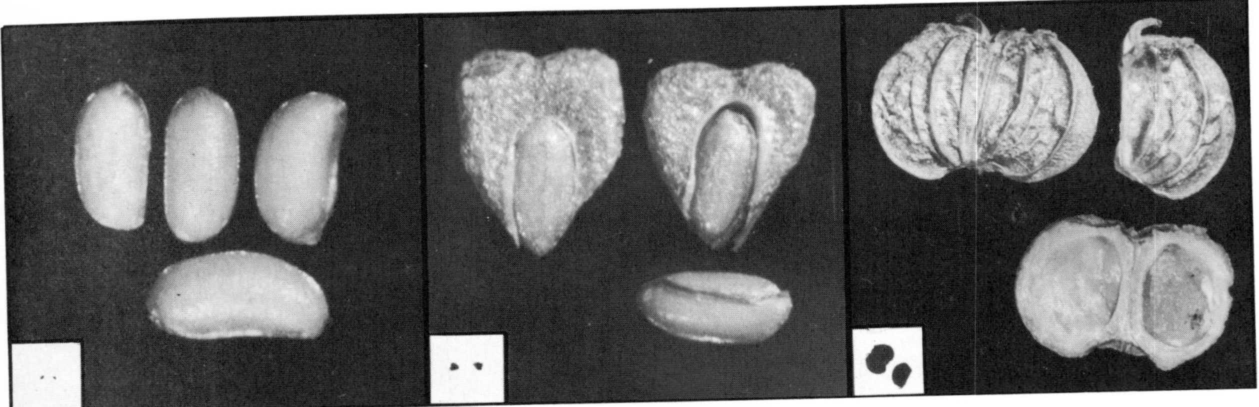

657. *Jussiaea decurrens*
Waterprimrose

658. *Jussiaea leptocarpa*
Waterprimrose

659. *Centella asiatica*
Centella

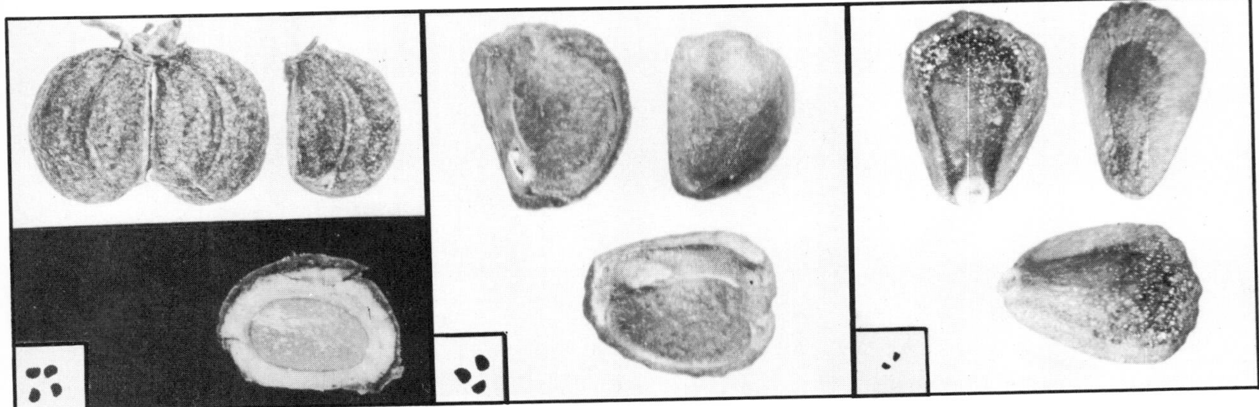

660. *Hydrocotyle umbellata*
Pennywort

661. *Heliotropium curassavicum*
Heliotrope

662. *Lycopus americanus*
Bugleweed

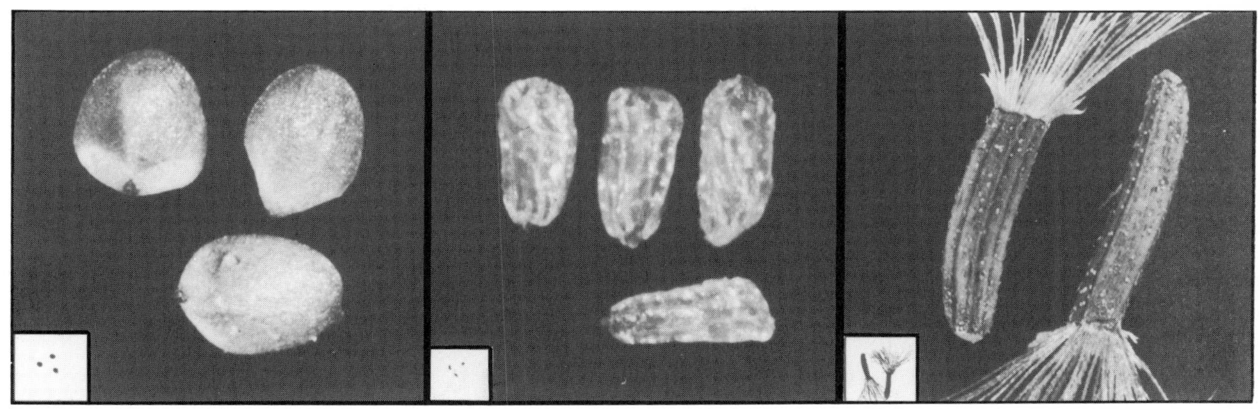

663. *Mentha canadensis*
Mint

664. *Bacopa monniera*
Waterhyssop

665. *Vernonia noveboracensis*
Ironweed

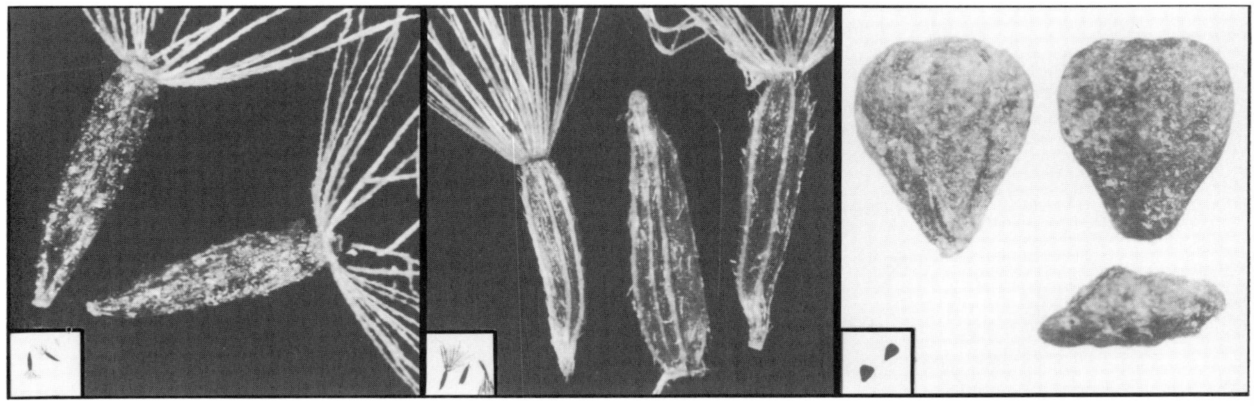

666. *Eupatorium perfoliatum*
Boneset

667. *Aster tenuifolius*
Aster

668. *Iva frutescens*
Sumpweed

669. *Iva xanthifolia*
Sumpweed

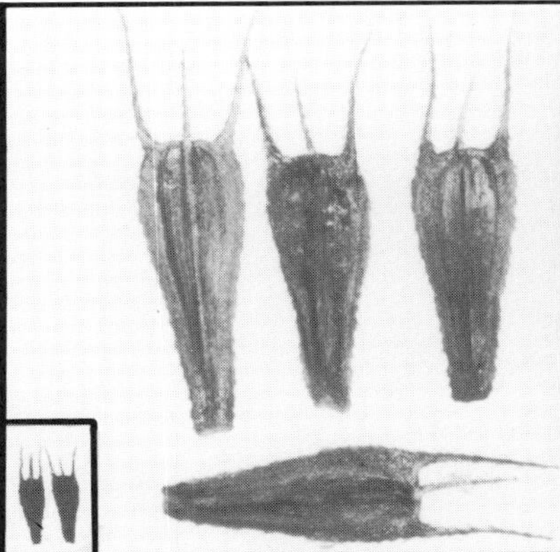

670. *Bidens comosa*
Sticktight

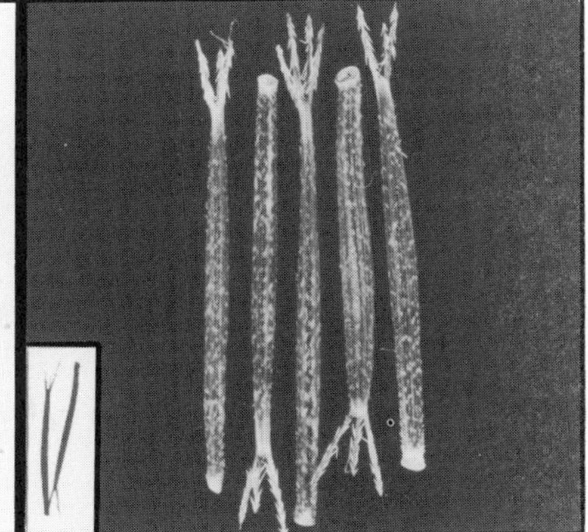

671. *Bidens bipinnata*
Sticktight

Woodlands

Woody Plants

Among the 132 species illustrated in this section are some woody plants which grow commonly in open areas, instead of being confined to woodlands. Only a few of the more common or more important trees and shrubs which are of interest to wildlife biologists and foresters have been included in the Woody Plants group. Actually, the seeds of woody plants justify a separate book, and one does exist. It is the well-illustrated Woody-Plant Seed Manual, United States Department of Agriculture Miscellaneous Publication 654, 1948.

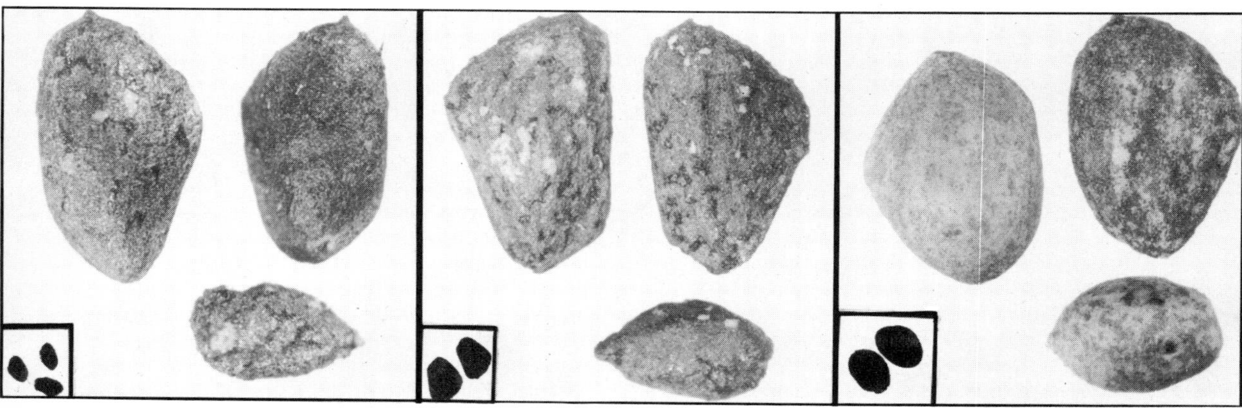

672. *Pinus rigida*
Pitch Pine

673. *Pinus taeda*
Loblolly Pine

674. *Pinus ponderosa*
Ponderosa Pine

675. *Pinus flexilis*
Limber Pine

676. *Pinus edulis*
Pinyon Pine

677. *Picea engelmannii*
Spruce

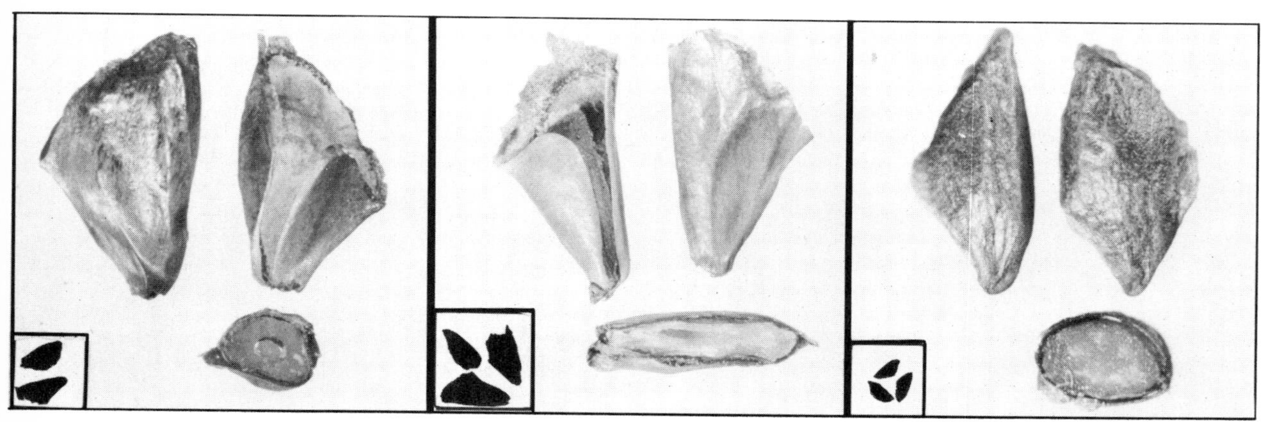

678. *Abies balsamea*
Fir

679. *Abies lasiocarpa*
Fir

680. *Tsuga canadensis*
Hemlock

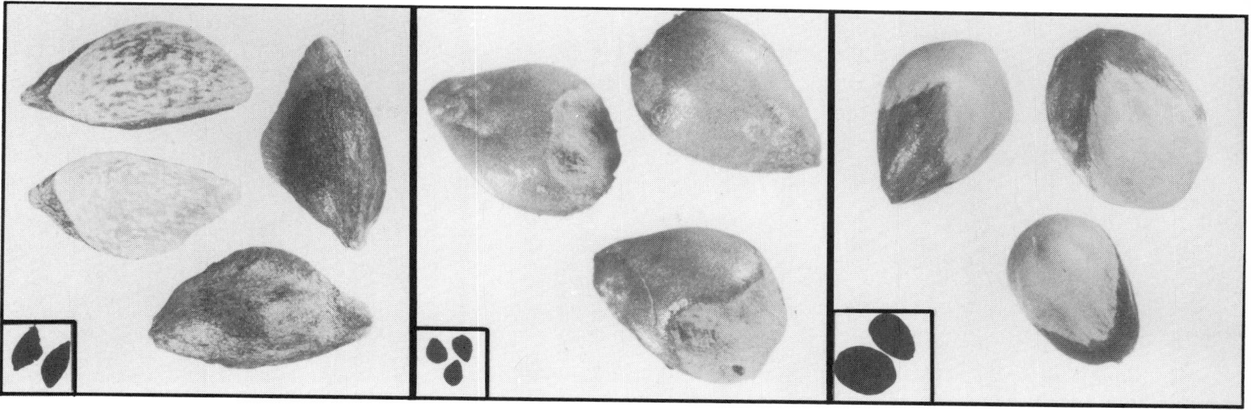

681. *Pseudotsuga taxifolia*
Douglasfir

682. *Juniperus virginiana*
Redcedar

683. *Juniperus utahensis*
Juniper

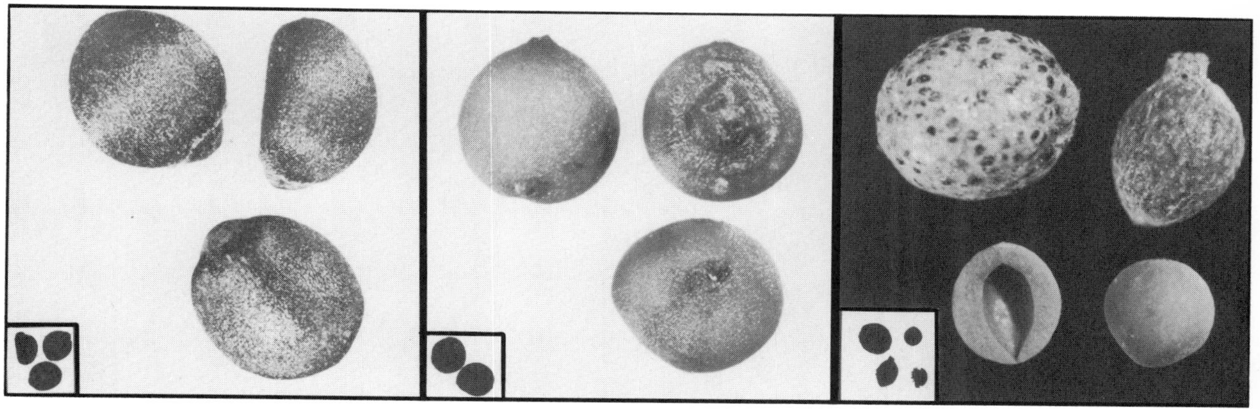

684. *Smilax glauca*
Greenbrier

685. *Smilax hispida*
Greenbrier

686. *Myrica pensylvanica*
Bayberry

687. *Carpinus caroliniana*
Hornbeam

688. *Ostrya virginiana*
Hop-hornbeam

689. *Betula lenta*
Birch

690. *Alnus serrulata*
Alder

691. *Fagus grandifolia*
Beech

692. *Quercus palustris*
Pin Oak

693. *Quercus falcata*
Spanish Oak

694. *Quercus marilandica*
Blackjack Oak

695. *Quercus emoryi*
Emory Oak

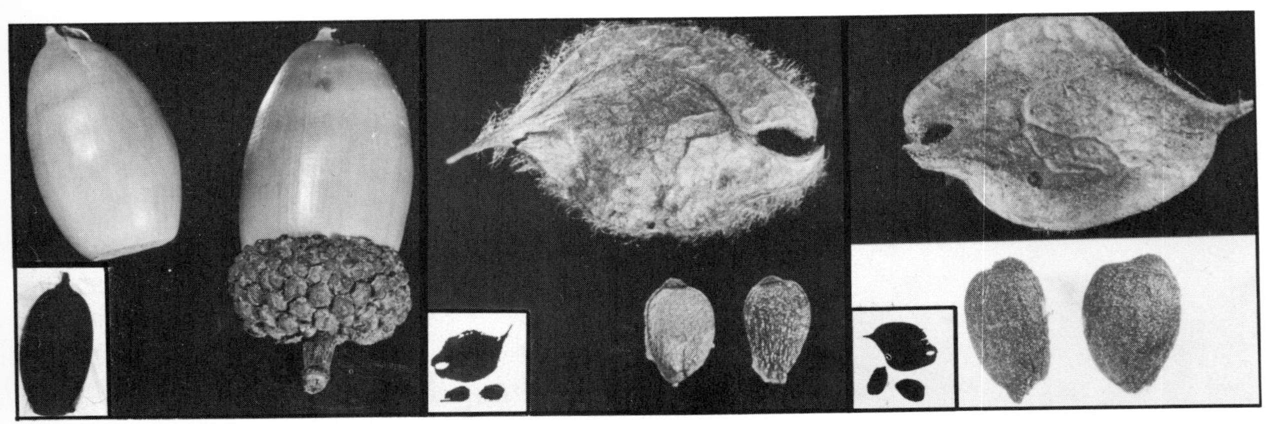

696. *Quercus alba*
White Oak

697. *Ulmus americana*
American Elm

698. *Ulmus crassifolia*
Cedar Elm

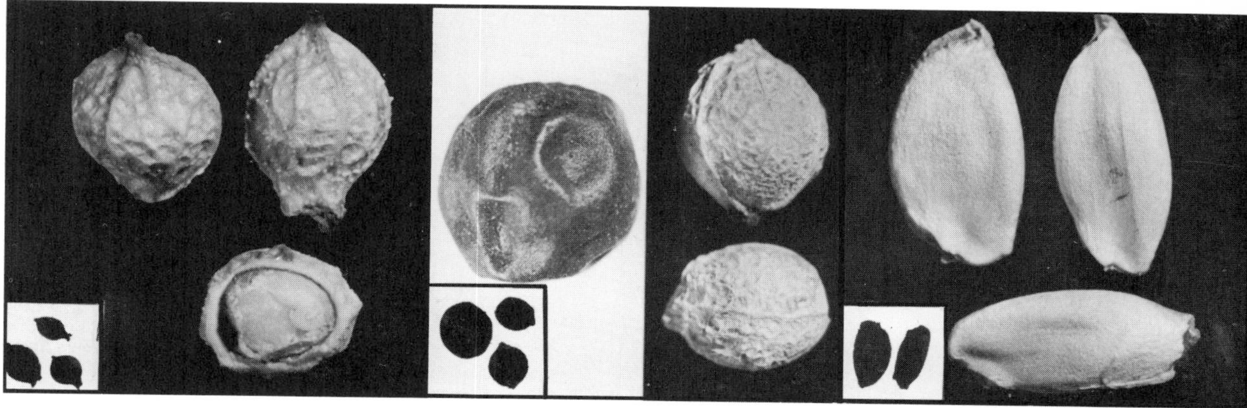

699. *Celtis pallida*
Hackberry

700. *Celtis mississippiensis*
Hackberry

701. *Maclura pomifera*
Osage-orange

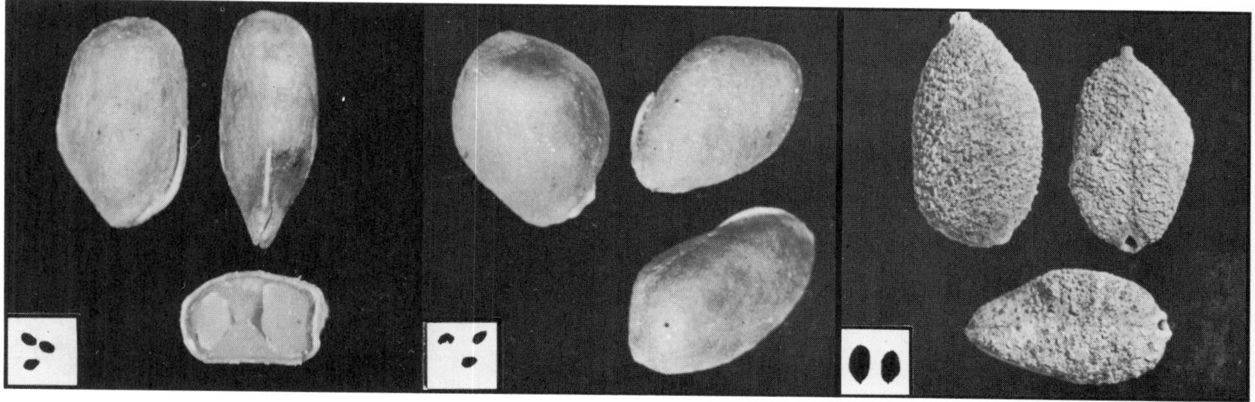

702. *Morus rubra*
Mulberry

703. *Morus alba*
Mulberry

704. *Liriodendron tulipifera*
Tuliptree

705. *Magnolia grandiflora*
Magnolia

706. *Magnolia virginiana*
Magnolia

707. *Sassafras officinale*
Sassafras

708. *Lindera aestivale*
Spicebush

709. *Liquidambar styraciflua*
Sweetgum

710. *Hamamelis virginiana*
Witch-hazel

711. *Platanus occidentalis*
Sycamore

712. *Malus pumila*
Apple

713. *Pyrus communis*
Pear

714. *Photinia arbutifolia*
Toyon

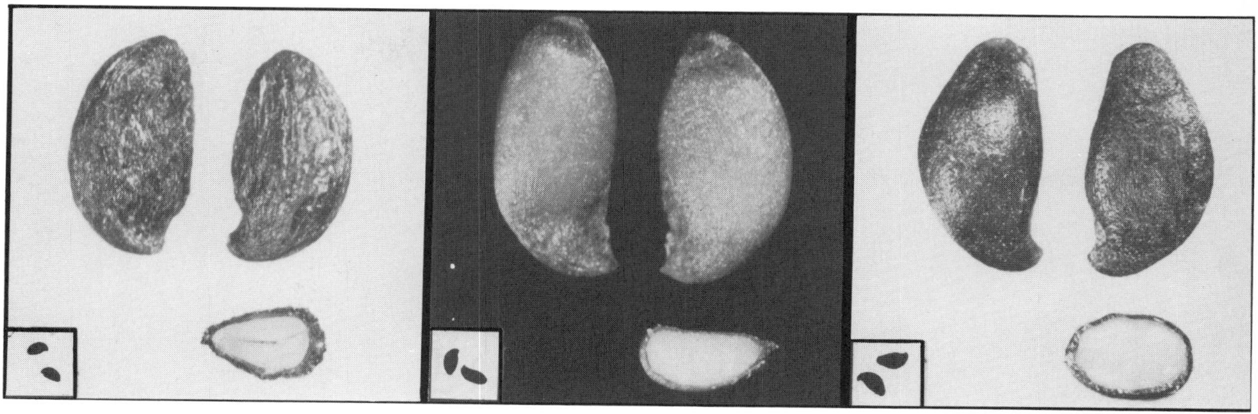

715. *Aronia arbutifolia*
Chokeberry

716. *Sorbus decora*
Mountain-ash

717. *Amelanchier canadensis*
Serviceberry

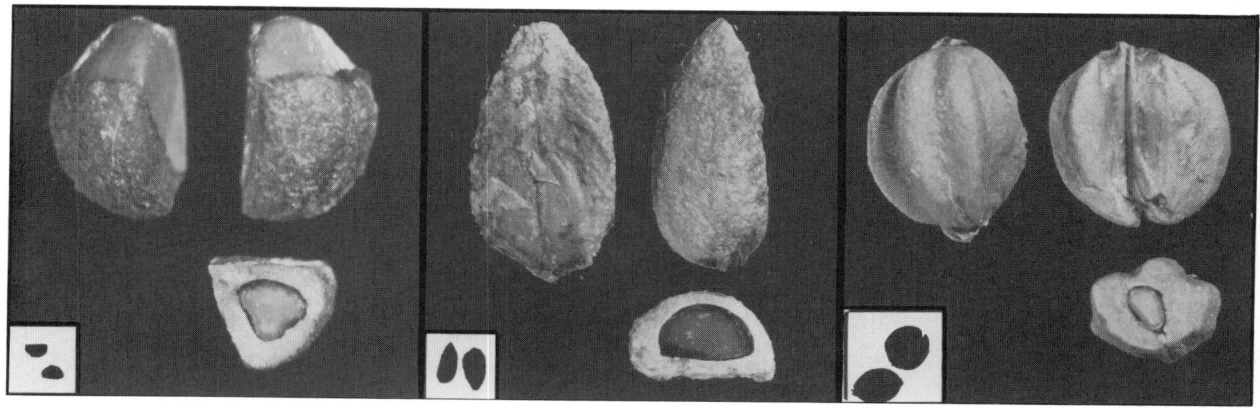

718. *Cotoneaster pyracantha*
Firethorn

719. *Crataegus apifolia*
Hawthorn

720. *Crataegus rotundifolia*
Hawthorn

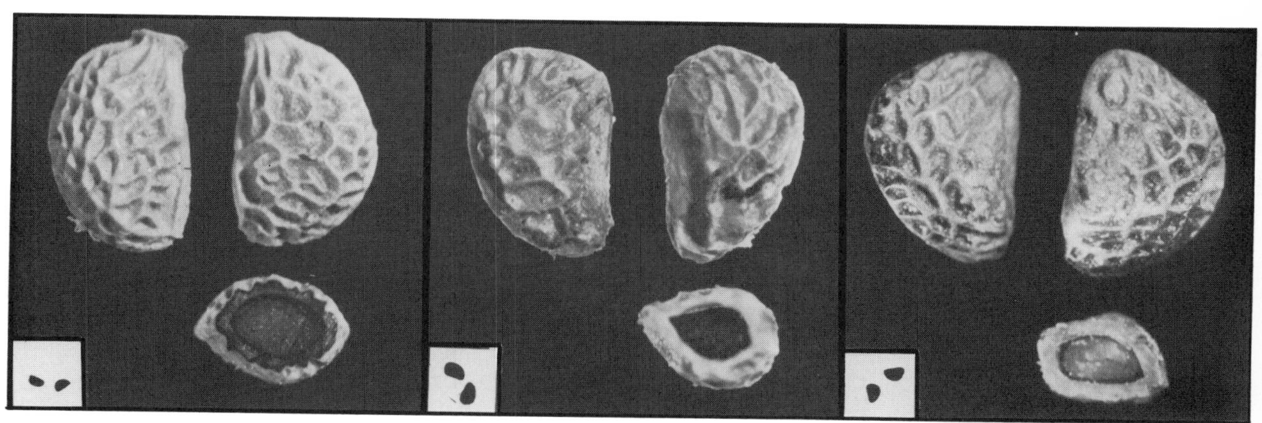

721. *Rubus odoratus*
Thimbleberry

722. *Rubus spectabilis*
Salmonberry

723. *Rubus occidentalis*
Raspberry

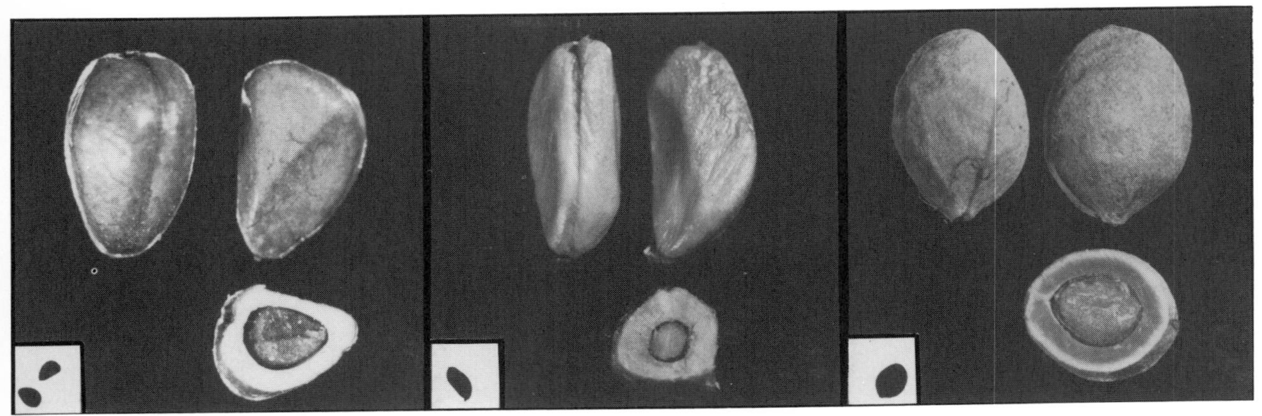

724. *Rosa palustris*
Rose

725. *Rosa woodsii*
Rose

726. *Prunus pensylvanica*
Pin Cherry

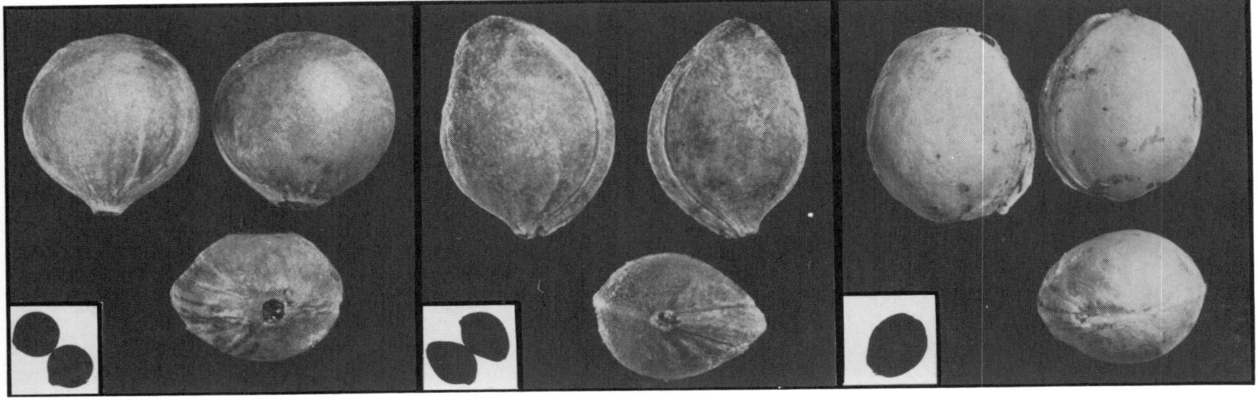

727. *Prunus serotina*
Black Cherry

728. *Prunus virginiana*
Choke Cherry

729. *Prunus cerasus*
Sour Cherry

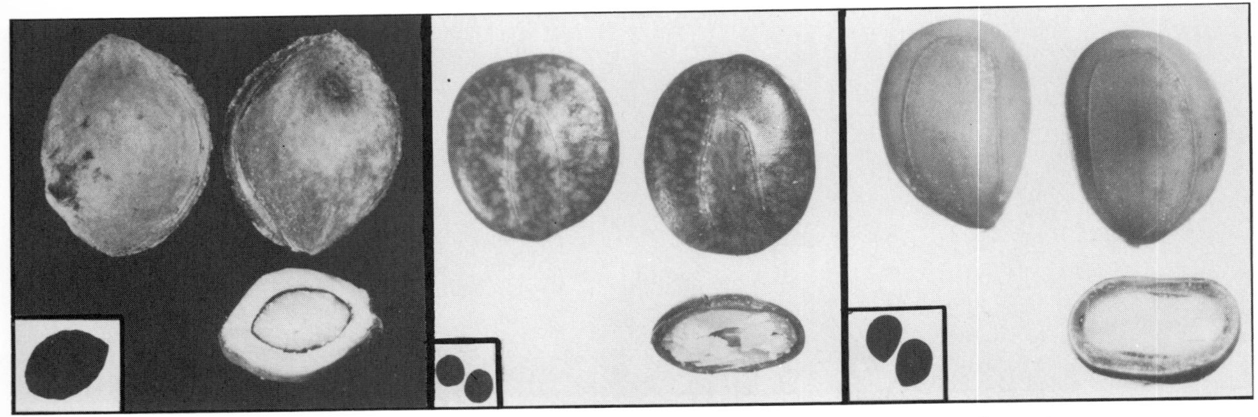

730. *Prunus americana*
Plum

731. *Acacia angustissima*
Acacia

732. *Acacia farnesiana*
Acacia

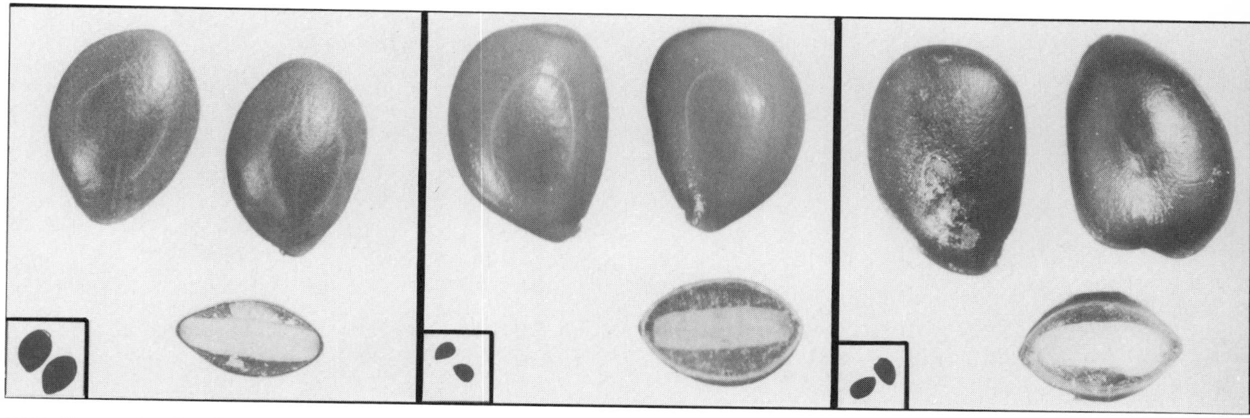

733. *Prosopis glandulosa*
Mesquite

734. *Prosopis pubescens*
Screwbean

735. *Mimosa biuncifera*
Mimosa

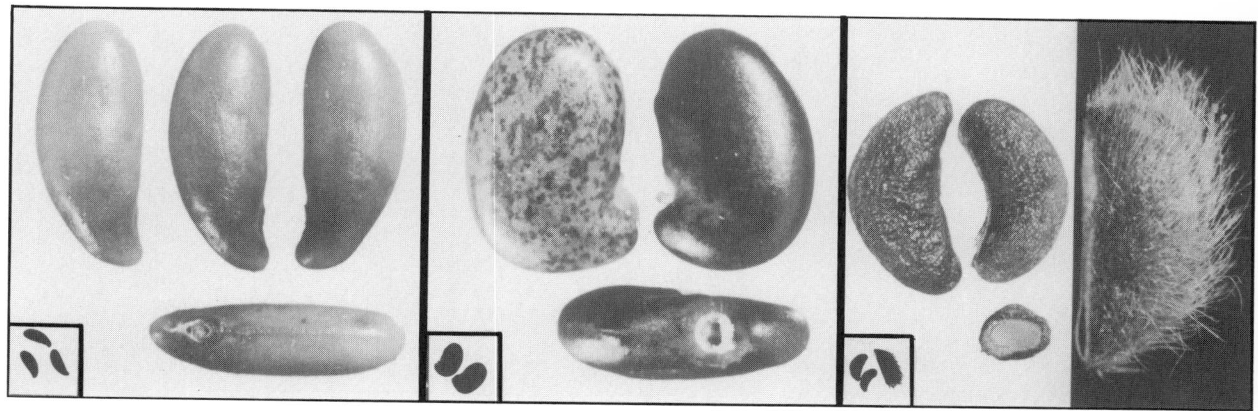

736. *Amorpha fruticosa*
Leadplant

737. *Robinia pseudoacacia*
Black Locust

738. *Larrea tridentata*
Creosote

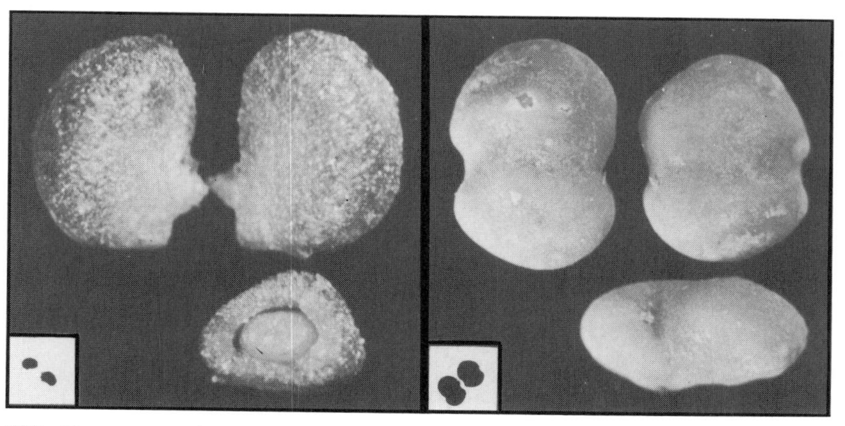

739. *Empetrum nigrum*
Crowberry

740. *Toxicodendron radicans*
Poison-ivy

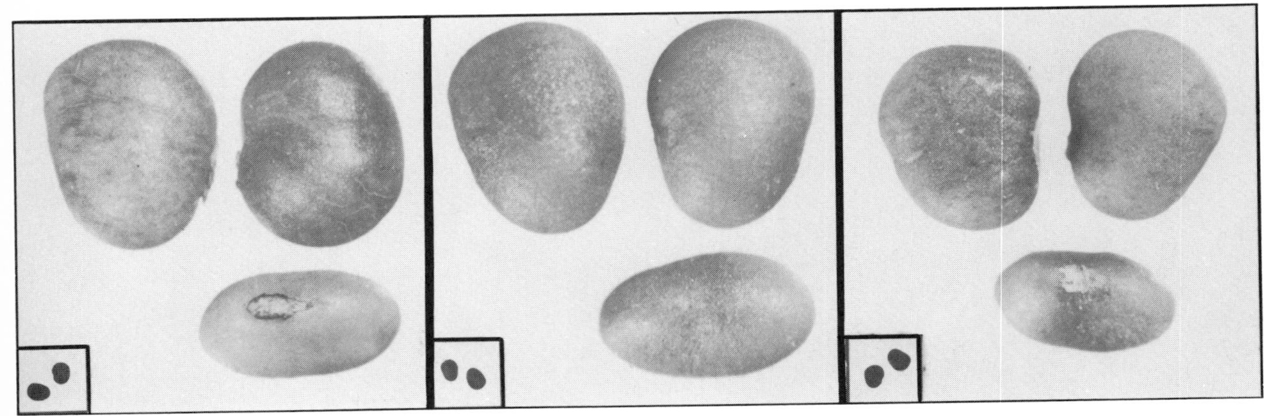

741. *Rhus copallina*
Dwarf Sumac

742. *Rhus glabra*
Smooth Sumac

743. *Rhus typhina*
Staghorn Sumac

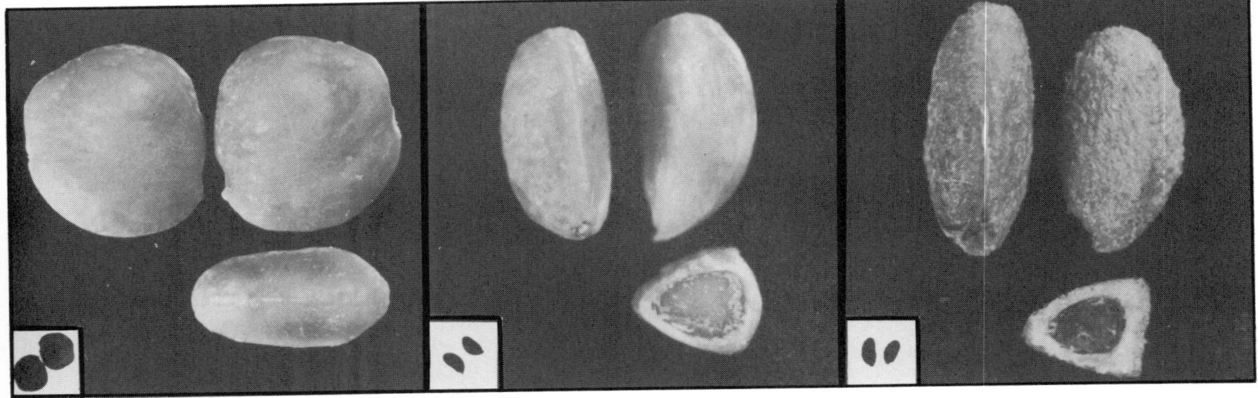

744. *Rhus trilobata*
Skunkbush

745. *Ilex verticillata*
Winterberry

746. *Ilex laevigata*
Winterberry

747. *Ilex glabra*
Inkberry

748. *Ilex decidua*
Possumhaw

749. *Ilex opaca*
Holly

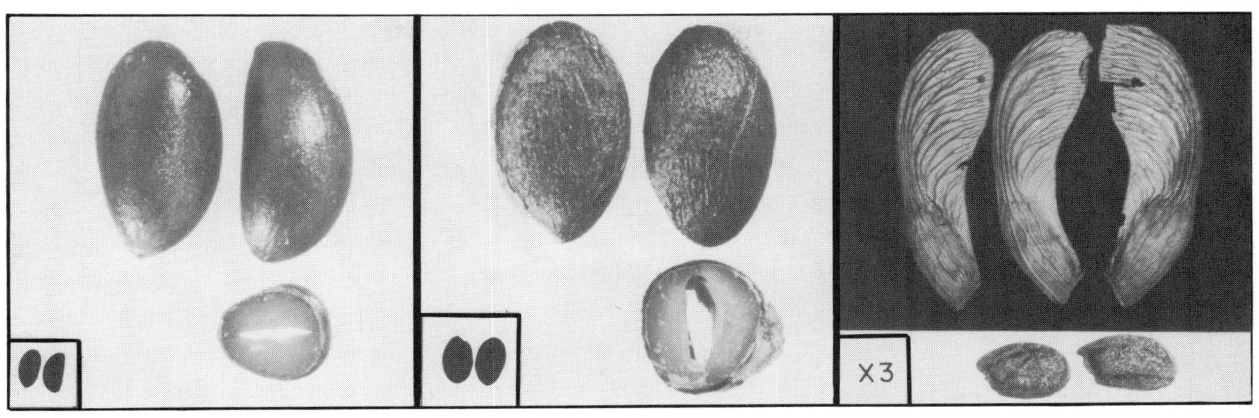

750. *Celastrus scandens*
Bittersweet

751. *Euonymus atropurpureus*
Wahoo

752. *Acer rubrum*
Red Maple

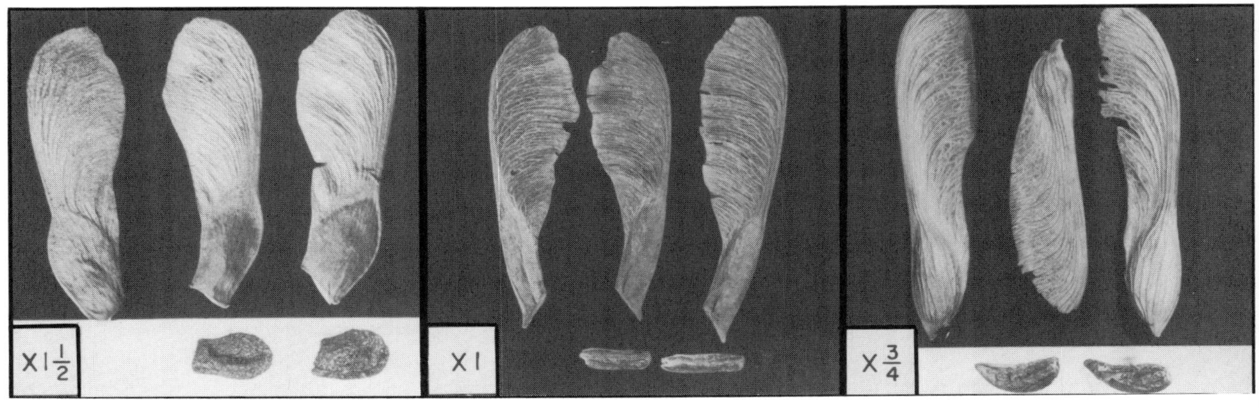

753. *Acer pensylvanicum*
Striped Maple

754. *Acer negundo*
Boxelder

755. *Acer saccharum*
Sugar Maple

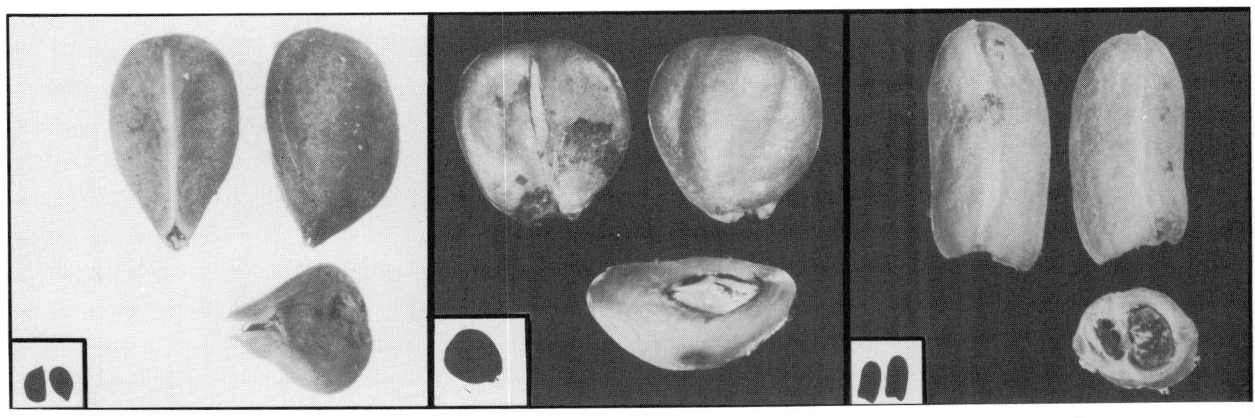

756. *Rhamnus cathartica*
Buckthorn

757. *Rhamnus californica*
Buckthorn

758. *Berchemia scandens*
Supplejack

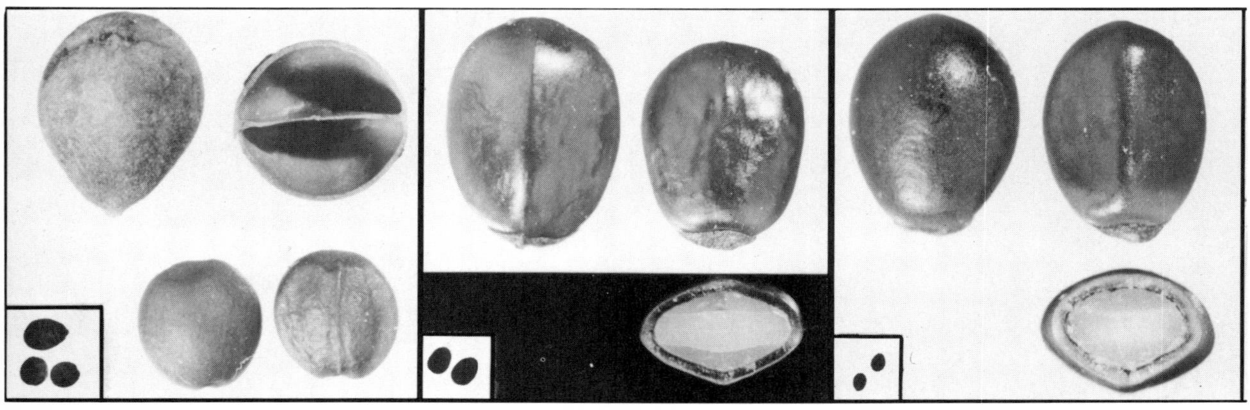

759. *Condalia obovata*
Condalia

760. *Ceanothus cuneatus*
Buckbrush

761. *Ceanothus americanus*
Jerseytea

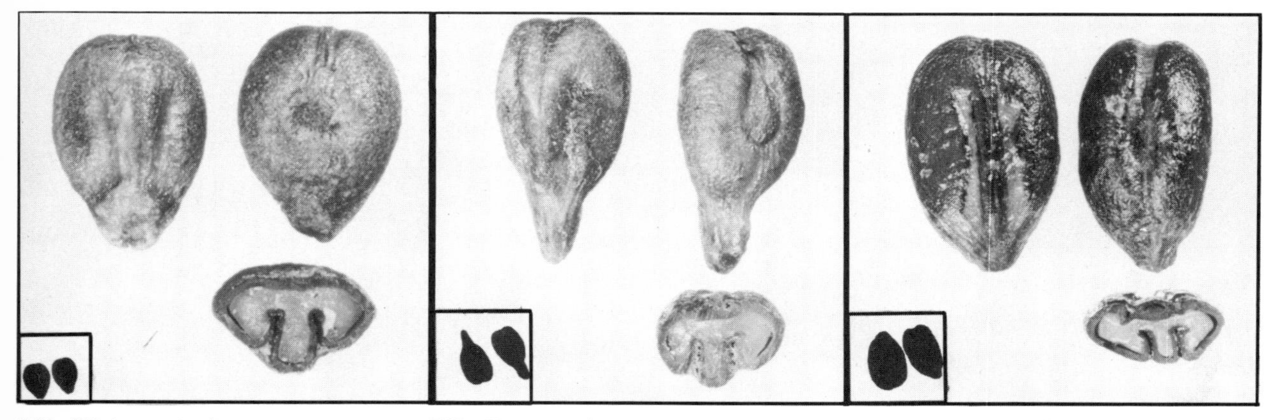

762. *Vitis aestivalis*
Grape

763. *Vitis vinifera*
Grape

764. *Vitis rotundifolia*
Muscadine

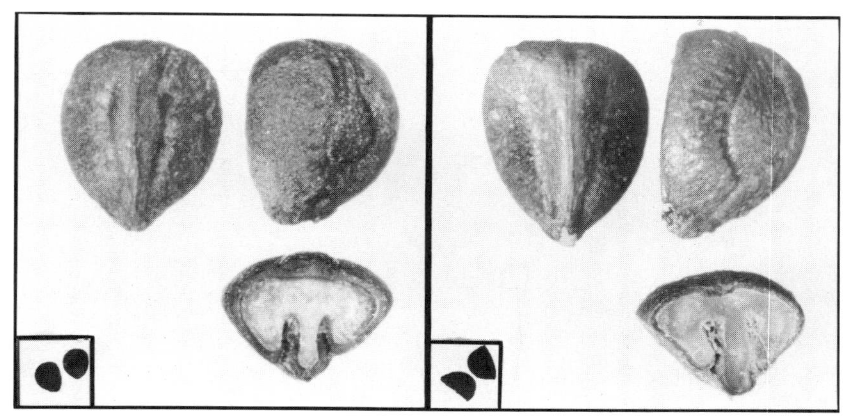

765. *Ampelopsis arborea*
Peppervine

766. *Parthenocissus quinquefolia*
Woodbine

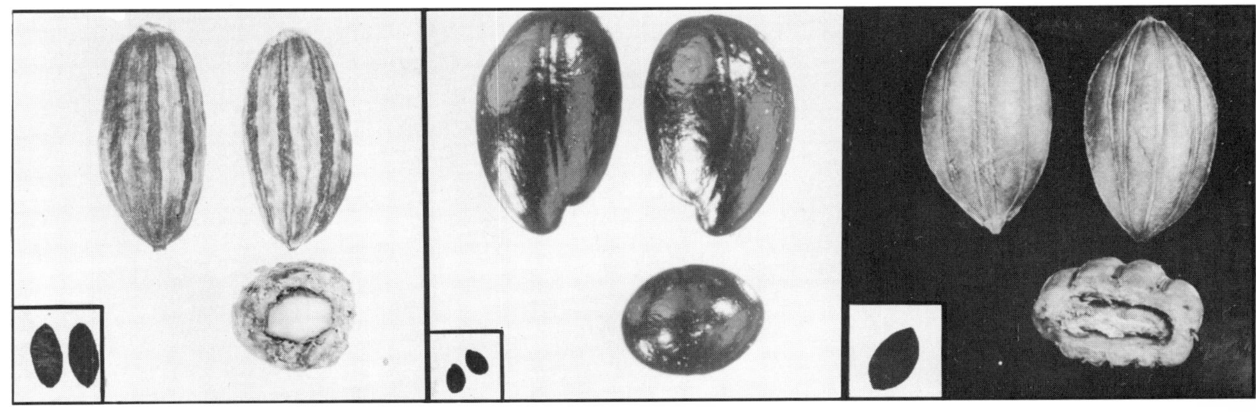

767. *Elaeagnus angustifolia*
Russianolive

768. *Shepherdia argentea*
Buffaloberry

769. *Nyssa biflora*
Blackgum

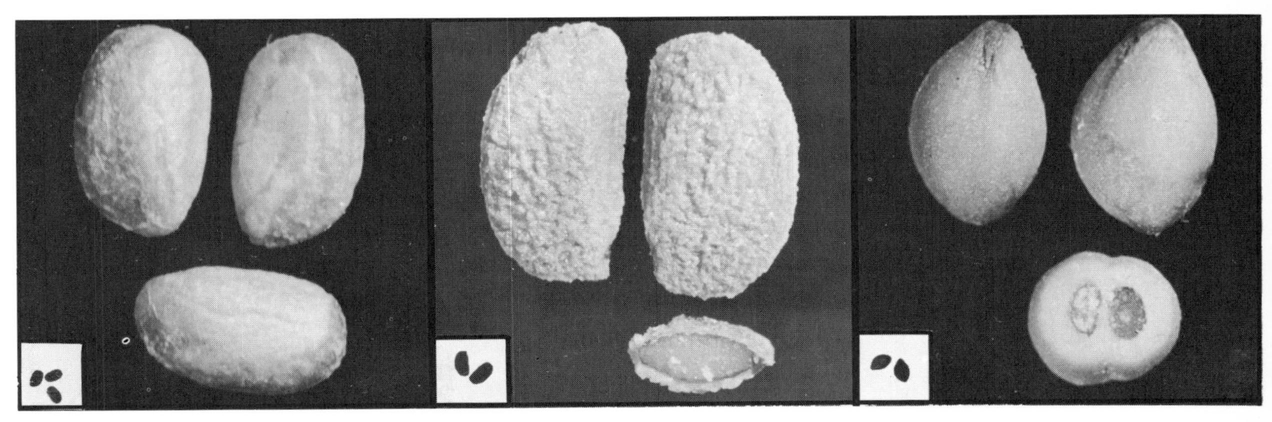

770. *Aralia spinosa*
Aralia

771. *Aralia racemosa*
Spikenard

772. *Cornus canadensis*
Bunchberry

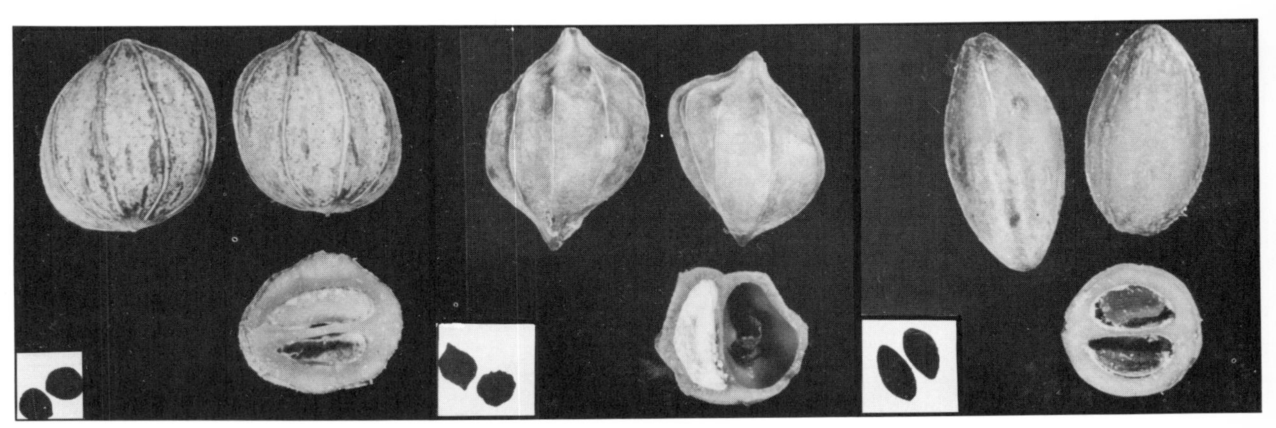

773. *Cornus paniculata*
Gray Dogwood

774. *Cornus amomum*
Silky Dogwood

775. *Cornus florida*
Flowering Dogwood

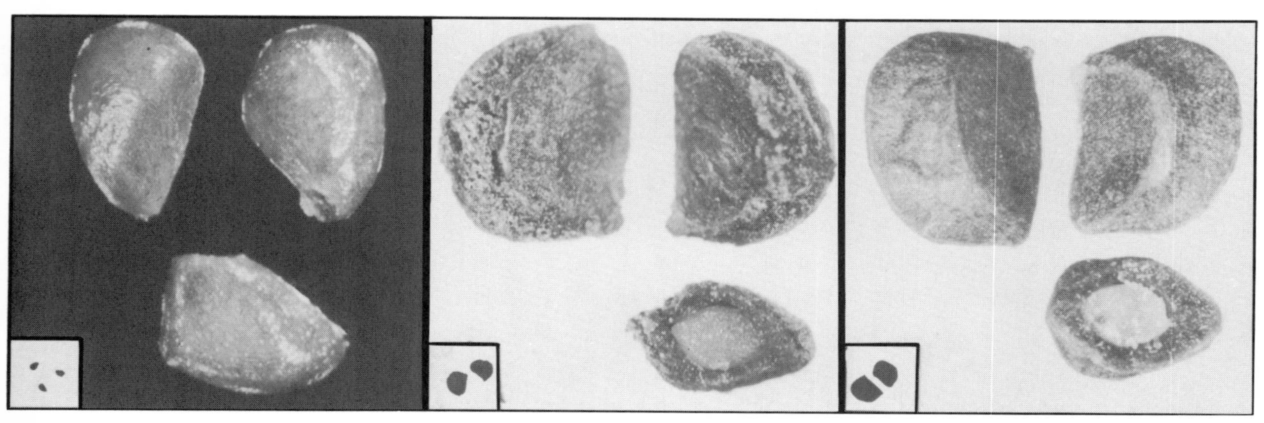

776. *Gaultheria procumbens* Wintergreen

777. *Arctostaphylos uva-ursi* Bearberry

778. *Arctostaphylos manzanita* Manzanita

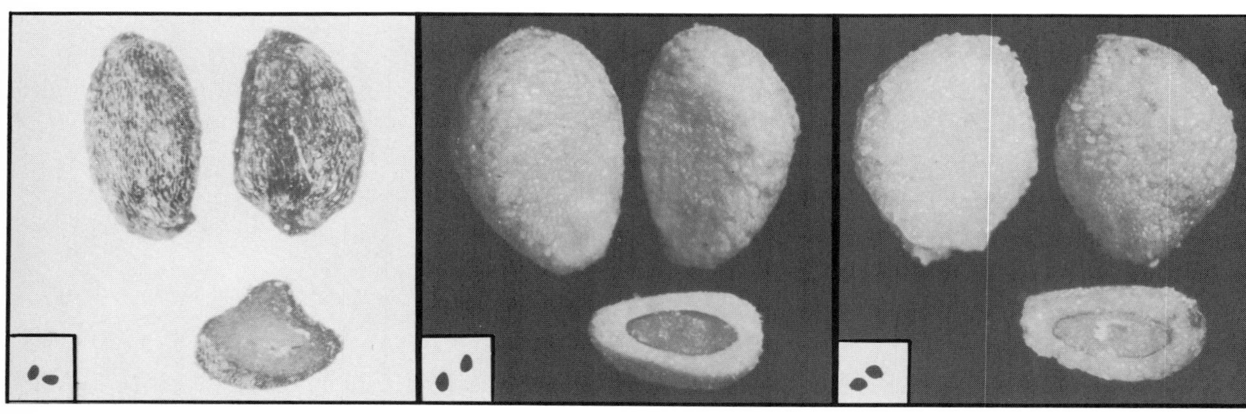

779. *Arbutus menziesii* Madrone

780. *Gaylussacia baccata* Huckleberry

781. *Gaylussacia frondosa* Dangleberry

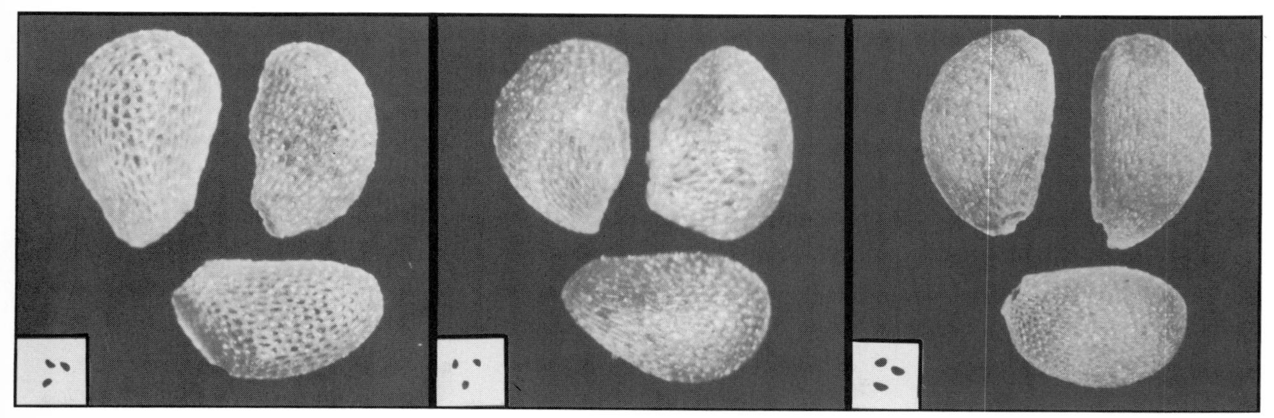

782. *Vaccinium vaccilans* Blueberry

783. *Vaccinium ovatum* Blueberry

784. *Vaccinium stamineum* Deerberry

785. *Bumelia lycioides*
Bumelia

786. *Diospyros virginiana*
Persimmon

787. *Forestiera angustifolia*
Waterprivet

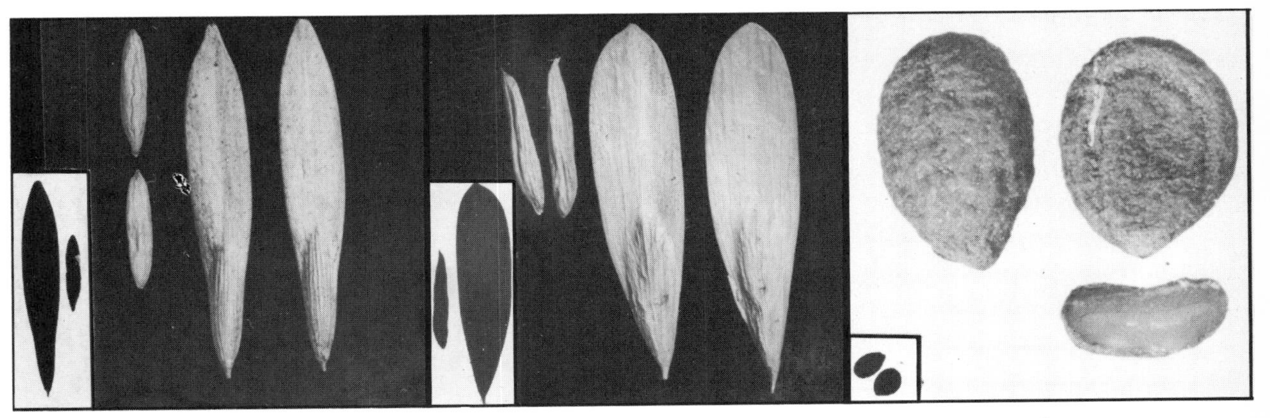

788. *Fraxinus americana*
Ash

789. *Fraxinus oregona*
Ash

790. *Ligustrum vulgare*
Privet

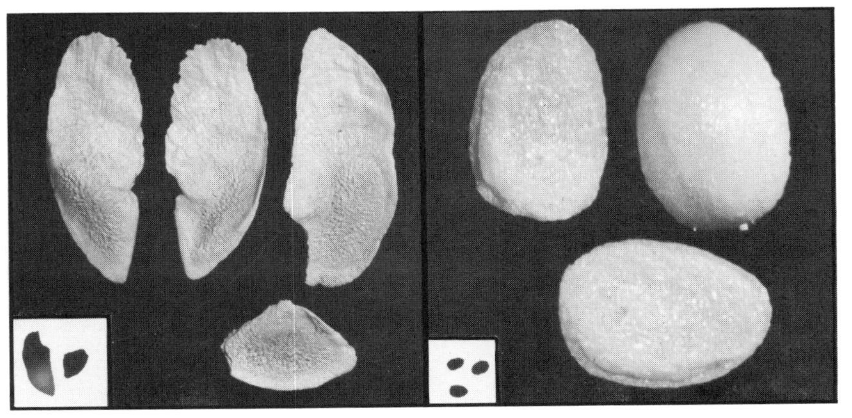

791. *Gelsemium sempervirens*
Jessamine

792. *Callicarpa americana*
Beautyberry

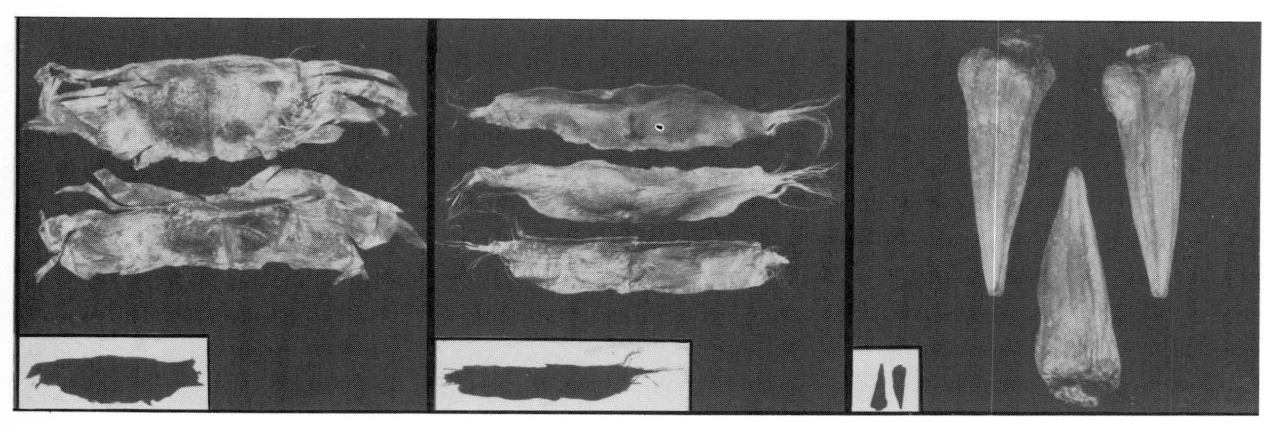

793. *Tecoma radicans*
Trumpetvine

794. *Catalpa bignonioides*
Catalpa

795. *Cephalanthus occidentalis*
Buttonbush

796. *Sambucus canadensis*
Elderberry

797. *Symphoricarpos orbiculatus*
Coralberry

798. *Symphoricarpos occidentalis*
Snowberry

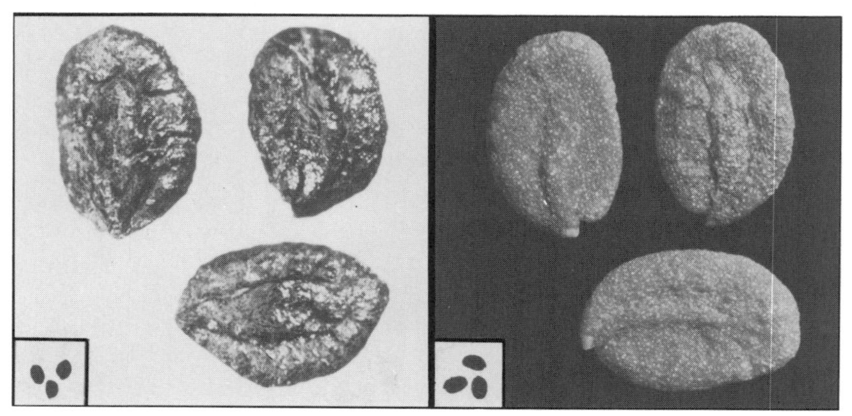

799. *Lonicera japonica*
Honeysuckle

800. *Lonicera tatarica*
Honeysuckle

801. *Viburnum dentatum*
Arrow-wood

802. *Viburnum acerifolium*
Viburnum

803. *Viburnum prunifolium*
Blackhaw

Woodlands

Nonwoody Plants

The twenty-one species of herbaceous or semiherbaceous plants represented here are common in woodlands in certain parts of the United States. Most of them are of value to various forms of woodland wildlife.

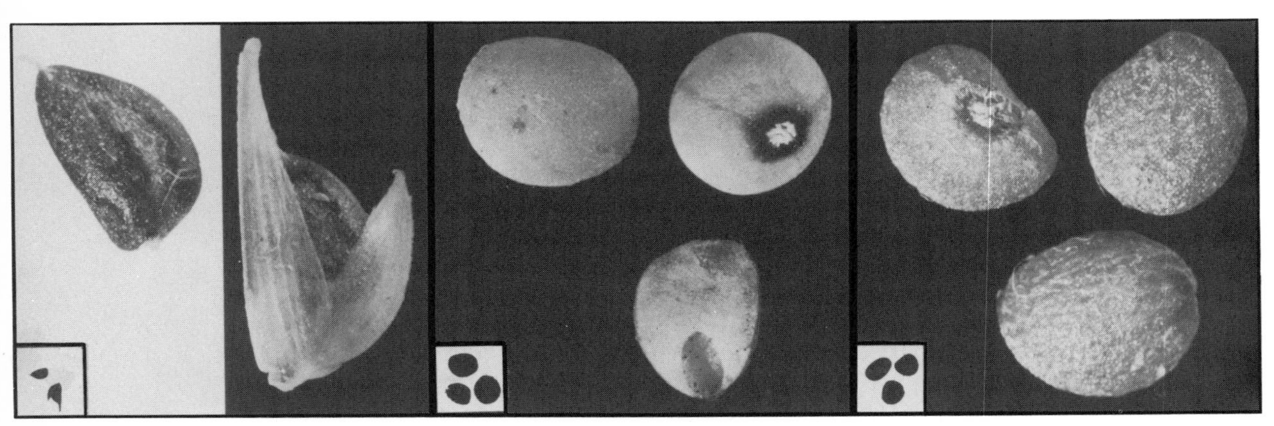

804. *Uniola laxa*
Uniola

805. *Smilacina racemosa*
False Solomonseal

806. *Polygonatum commutatum*
Solomonseal

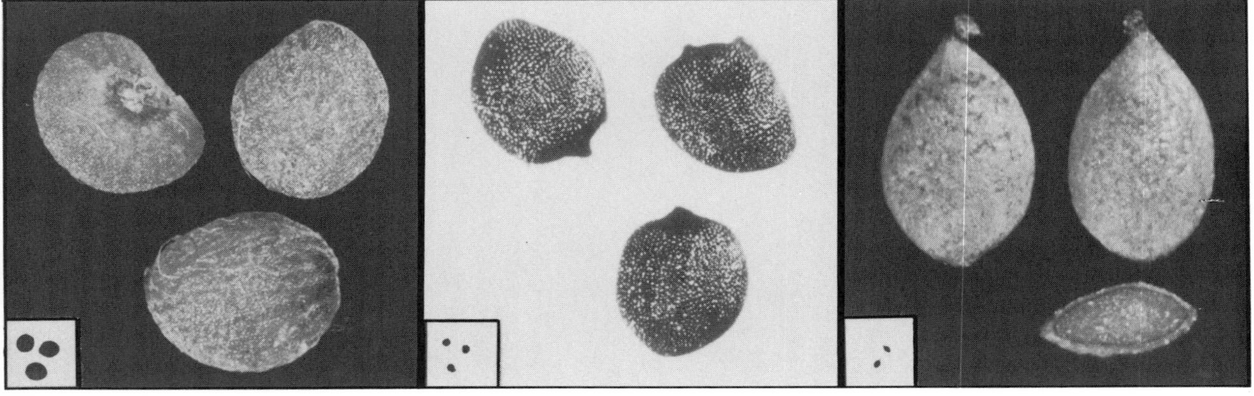

807. *Maianthemum canadense*
Beadruby

808. *Hypoxis hirsuta*
Yellow Stargrass

809. *Urtica lyallii*
Nettle

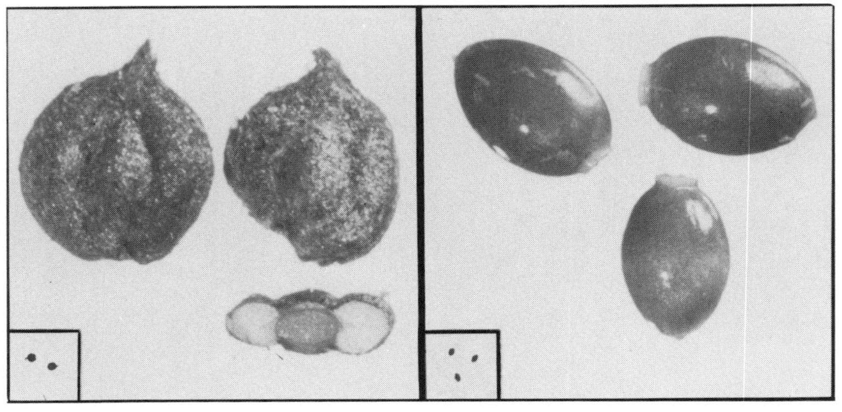

810. *Boehmeria cylindrica*
Falsenettle

811. *Parietaria pensylvanica*
Pellitory

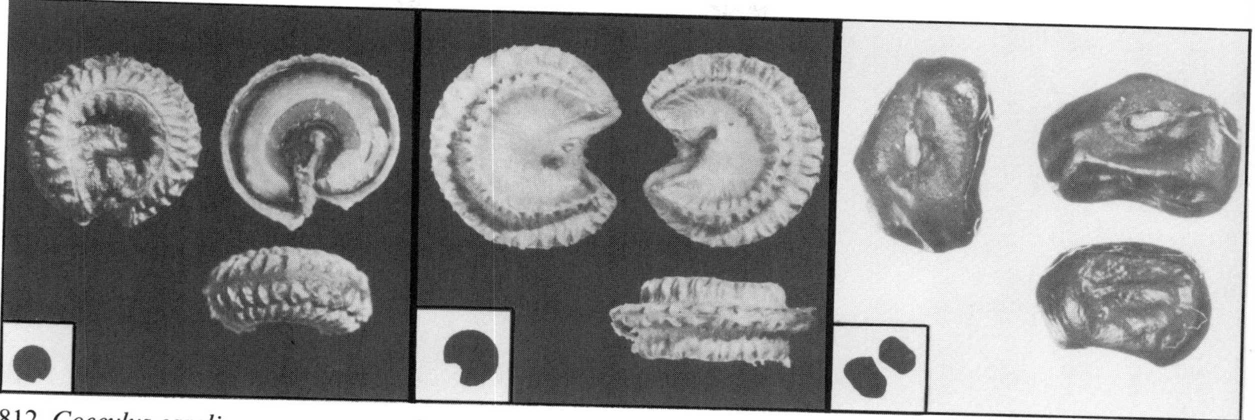

812. *Cocculus carolinus*
Snailseed

813. *Menispermum canadense*
Moonseed

814. *Apios tuberosa*
Groundnut

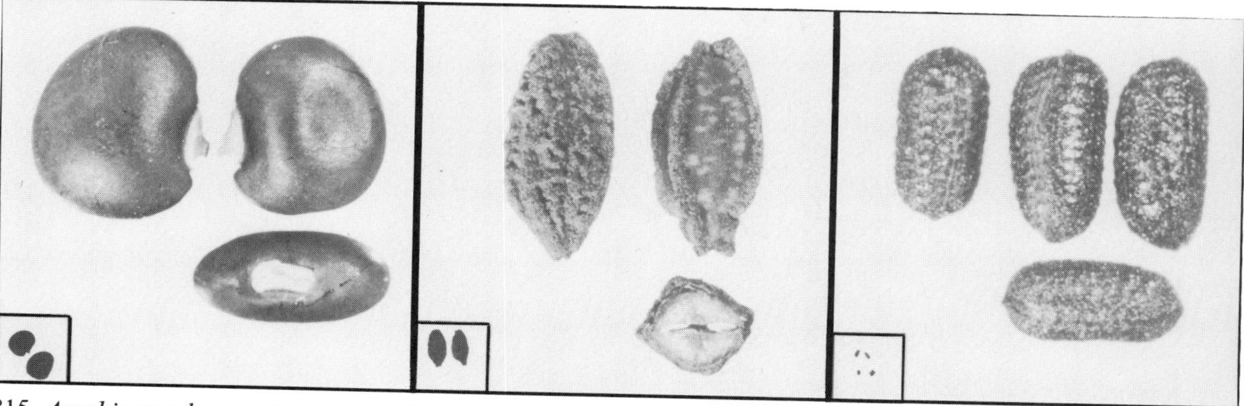

815. *Amphicarpa bracteata*
Hogpeanut

816. *Impatiens biflora*
Jewelweed

817. *Hypericum perforatum*
St. Johnswort

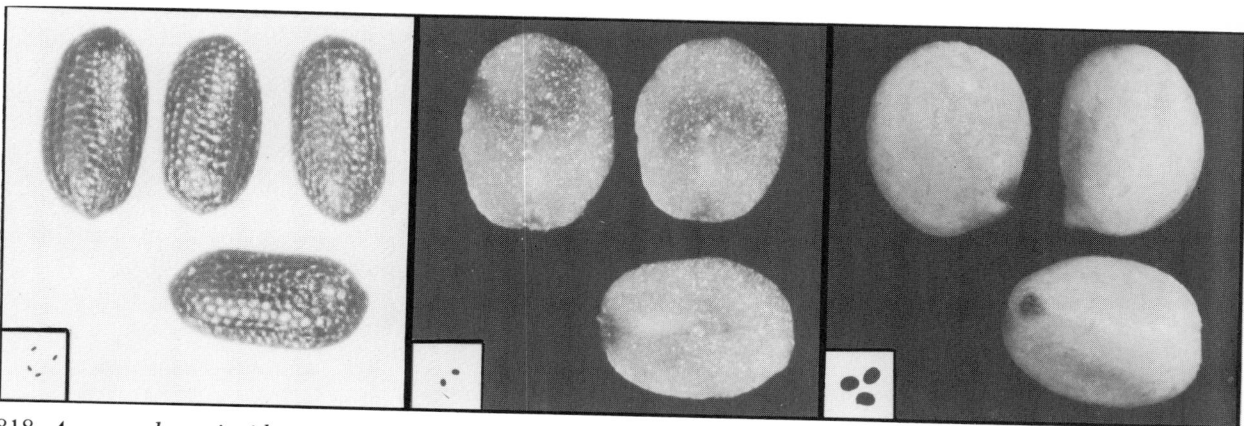

818. *Ascyrum hypericoides*
St. Andrewscross

819. *Veronica officinalis*
Speedwell

820. *Mitchella repens*
Partridgeberry

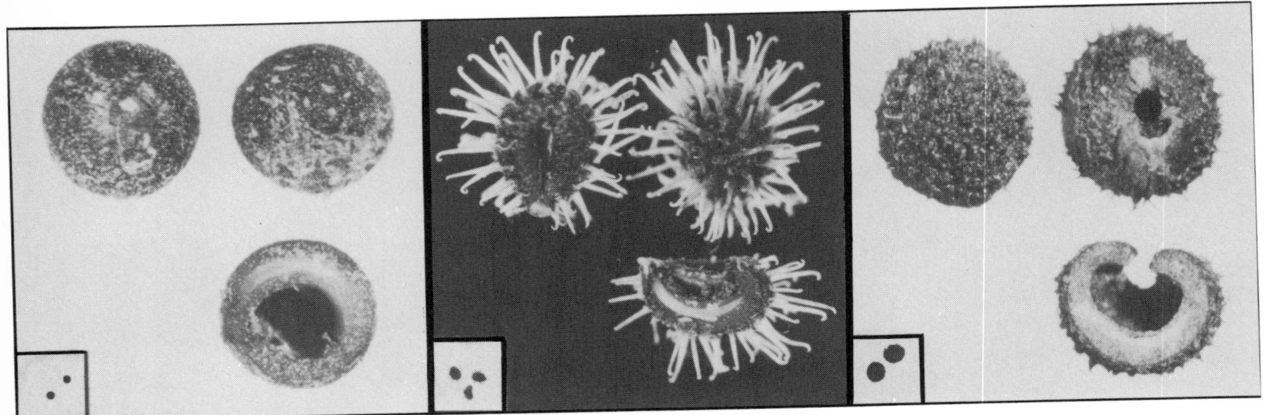

821. *Galium trifidum*
Bedstraw

822. *Galium aparine*
Bedstraw

823. *Galium boreale*
Bedstraw

824. *Aster cordifolius*
Aster

Identification Clues

This part of the manual points out and describes characteristics which are useful in distinguishing seeds of many important families and genera of the plant kingdom. With the exception of groups such as the pondweeds (*Potamogeton*) and smartweeds (*Polygonum*), species are mentioned or illustrated chiefly to typify genera or families in which they belong. Actually, it is not feasible or safe to try to identify the species of seeds in a major proportion of genera.

Identification features used include significant internal characteristics as well as external ones. Gross internal morphology has proved particularly helpful in identifying the family or genus of a seed. This approach helps an investigator to start with the right major group. Especially useful internal clues include the amount, position, and nature of the endosperm, if any, or the absence of endosperm; the size, shape, cotyledon development, and position of the embryo; and the texture, thickness, and inner markings of the seed wall.

Important clues are emphasized by **boldface** type.

The treatment is in systematic order, from the Pinaceae to the Compositae. In instances wherein only one genus of a family is discussed, family descriptions are omitted. On occasion, two or even three similar and closely related genera are described jointly for more effective comparison and distinction.

The plate references, after generic and common names in the side headings, are to the appropriate photographs in the systematic sections of the first part of the manual, Seed Photographs. If the reader wishes to see the duplicate plates in Weeds, Arranged by Physical Features, in conjunction with other plates showing similar features, he can locate them through the Index. Since there are no generic side heads for the Gramineae, for reasons given in the text and in the Introduction, the grass genera pictured in Seed Photographs are listed in a footnote (p. 133, n. 3), so that their respective plate numbers also can readily be obtained from the Index. As is mentioned in the Introduction, the common names of the plants are, with relatively few modifications, those given in Standardized Plant Names, by Harlan P. Kelsey and William A. Dayton (2d ed., Harrisburg, Pa., 1942).

PINACEAE Pine Family

An important identification aid in this family is the **papery, brownish inner wall (tegmen or nucellus)**, similar to the one which helps distinguish seeds of the **hornbeams** (*Carpinus* and *Ostrya*) and beech family from other families and genera. Seeds are winged in most genera. In pines, spruces, hemlocks, and other genera having delicate wings which detach easily, the wing tissue shows distinctive **wavy, fingerprint-like lines running lengthwise** (fig. 2). The embryo is linear in oily-fleshy endosperm, and the **cotyledons** (fig. 3) **are commonly more than 2,** except in *Thuja, Chamaecyparis,* and *Libocedrus.*

PINUS Pine (pls. 672–676)

Seed length (exclusive of wings) ranges from 3 to 24 mm., and wall thickness from 0.2 to 1.5 mm. **Fingerprint-like markings** (coarser and less distinct than those on the wings) are on the wall's innder surface. Cotyledons vary from 3 to 18.

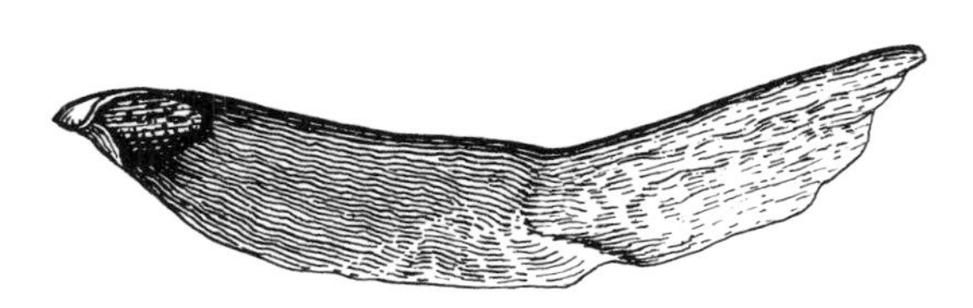

Fig. 2. Pine seed (*Pinus taeda*)

Fig. 3. Pine embryo

Fig. 4. *Pinus taeda* X 2.5

PICEA Spruce (pl. 677)

Biconvex; 2.5–4 mm. long, exclusive of wings; resin pits lacking; the wall's inner surface has **fine, short, wavy, lengthwise lines.**

ABIES Fir (pls. 678, 679)

Seeds larger than in spruces; without wings, they vary from 5 to 17 mm. in length; some faces of seeds are **flattish or plane, and shiny;** inner surface of wall shows **fine, wavy markings that run diagonally** or nearly crosswise.

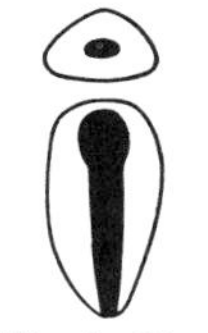

Fig. 5. *Picea engelmannii* X 5.5

Fig. 6. *Abies lasiocarpa* X 2.5

TSUGA Hemlock (pl. 680)

Exclusive of wings, seeds of different species are 3.5–4 mm. long. Though similar to spruce seeds in size, they are distinct in their **protruding resin chambers,** and in obscure markings on the inner face of the wall.

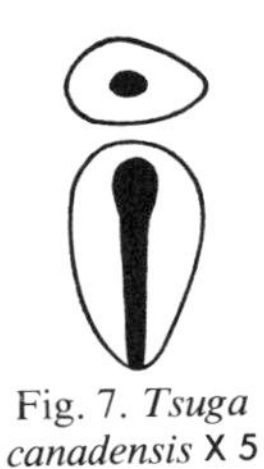

Fig. 7. *Tsuga canadensis* X 5

Fig. 8. *Pseudotsuga taxifolia* X 3.5

PSEUDOTSUGA Douglasfir (pl. 681)

Biconvex; 6–12 mm. long, excluding wings, in our two species; surface **demarked clearly into two zones,** one brownish, with distinct cellular markings, and the other whitish, with obscure cellular markings.

JUNIPERUS Juniper, Redcedar (pls. 682, 683)

Ovoid; 4–8 mm. long; with prominent **roundish, scarlike area** at broad end, and usually pointed at the other. Resin pits evident on some. Wall woody, its thickness ranging from 0.3 to 1.3 mm. in various species and differing considerably in different parts of the same seed. Seeds borne in berry-like fruits.

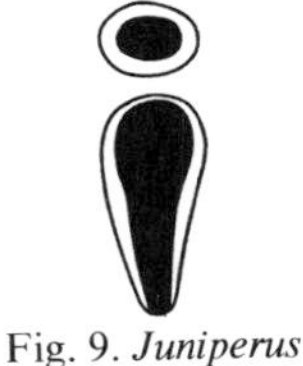

Fig. 9. *Juniperus virginiana* X 4

TYPHACEAE Cattail Family

TYPHA Cattail (pl. 620)

Seeds **minute,** rodlike, 1–2 mm. long, with a small, conical nipple on the truncate end. The very thin (0.02 mm.) wall is roughened by granule-like papillae and cellular reticulations. Fruits air-borne by characteristic **whorls of fine hairs** (fig. 10). Embryo linear.

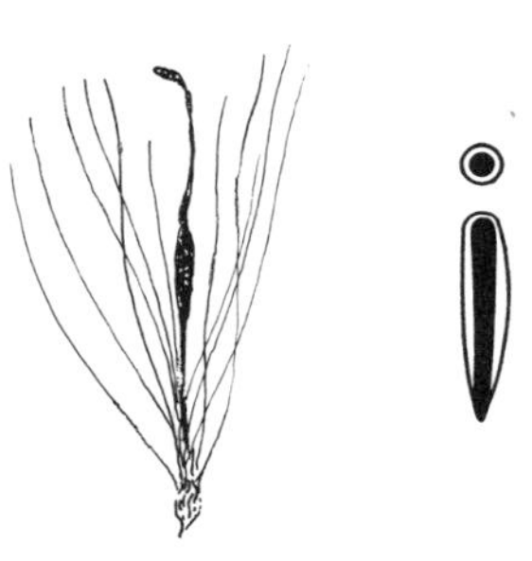

Fig. 10. *Typha latifolia.* Diagram (right) X 14

SPARGANIACEAE Burreed Family

SPARGANIUM Burreed (pls. 621, 622)

Shape and size vary in different species; usually ellipsoid to ovoid; **pithy tissues enclose a woody central core** and in some species constitute a bulbous area in the upper half of the seed (nutlet). Commonly there is a slight constriction in the waistlike zone below the pithy tissue. *S. eurycarpum* (fig. 11) is distinct in being **angled-obovate,** with broad shoulders, and is unusually large, varying from 6 to 10 mm. in length and from 6 to 8 mm. in width. Often it has 2 seed chambers.

Other burreed species are ellipsoid to ovoid in outline and are difficult to distinguish. In all, the **slender embryo** extends like an axle through a **core of white, starchy endosperm.**

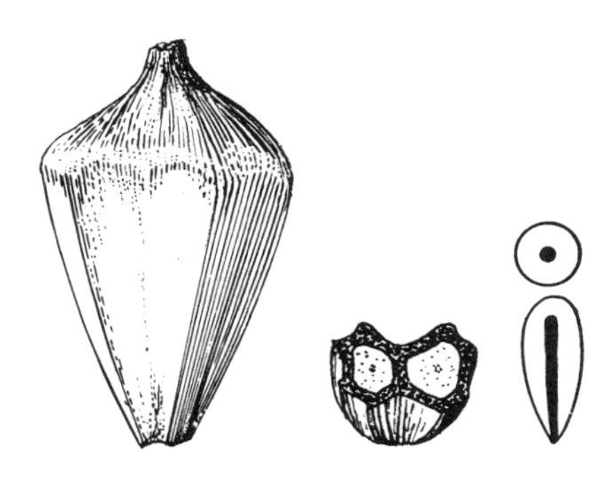

Fig. 11. *Sparganium eurycarpum.* Diagram (right) X 3.5

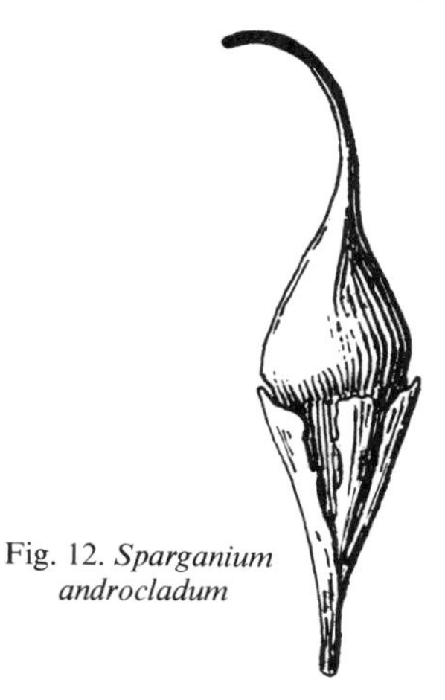

Fig. 12. *Sparganium androcladum*

NAJADACEAE Pondweed Family

Seed (nutlet) shapes, sizes, and surfaces are diverse among genera of this family of aquatics. The only consistent family characteristic significant for identification is the **lack of endosperm.**

POTAMOGETON Pondweed (pls. 529–539)

Diagnostic features of pondweed seeds with their outer covering removed and also with cut sections are shown in figure 13. The characteristic **earlike appearance** caused by the curved or coiled embryo is evident in the cut sections. Figures 14–16[1] illustrate (X 10) the seeds of twenty-one important species of pondweeds with their seed coverings removed, in approximate order of size, small to large.

---

[1] From Identifying Pondweed Seeds Eaten by Ducks, by Alexander C. Martin, Jour. Wildl. Mgmt., 15:253–258, 1951.

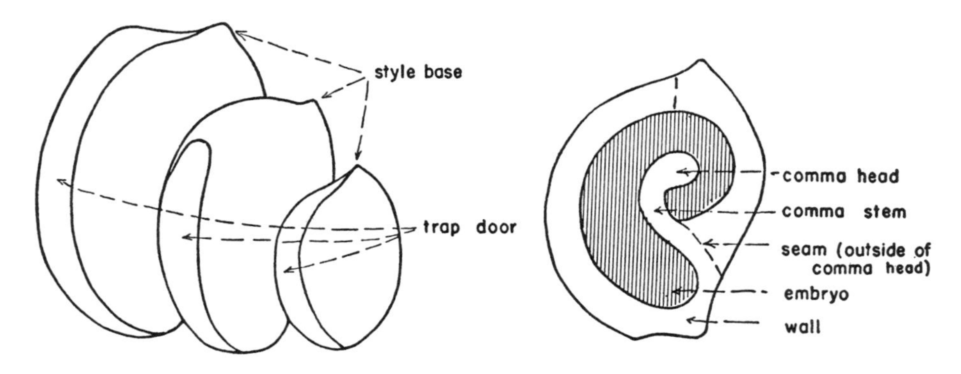

Fig. 13. External and internal features of pondweed seeds

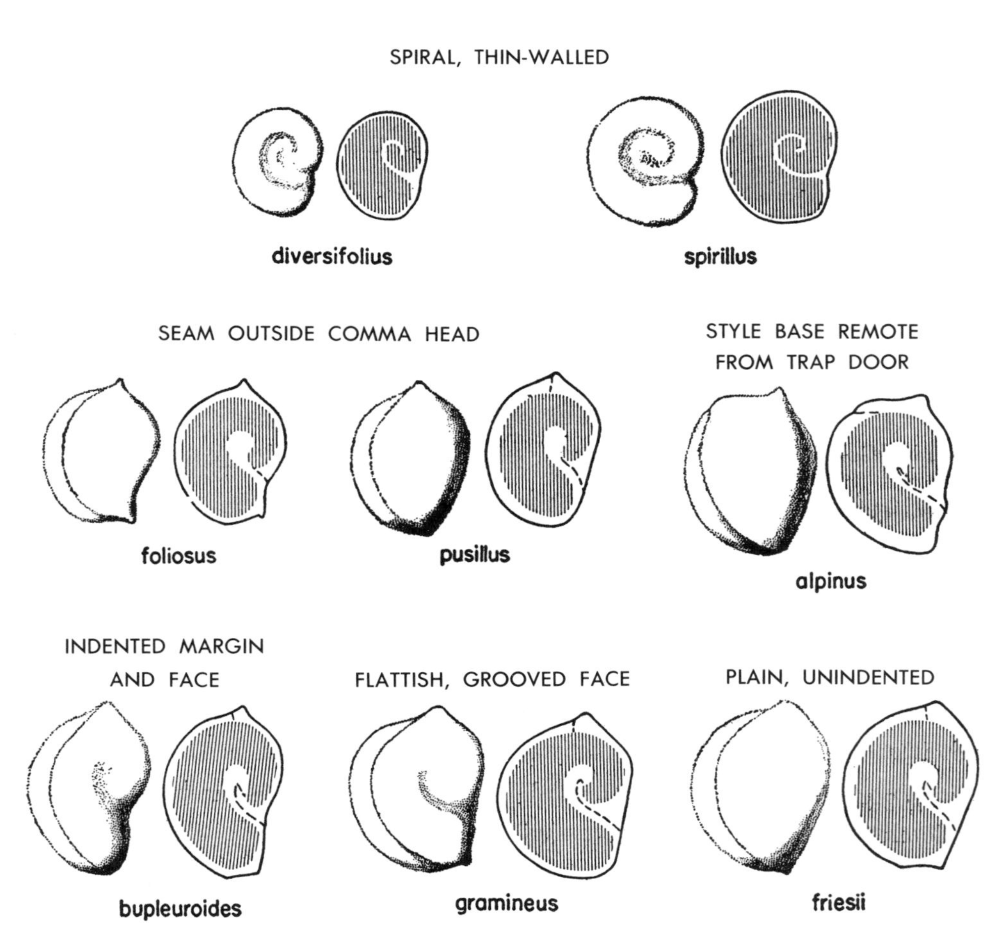

Fig. 14. Pondweed (*Potamogeton*) seeds

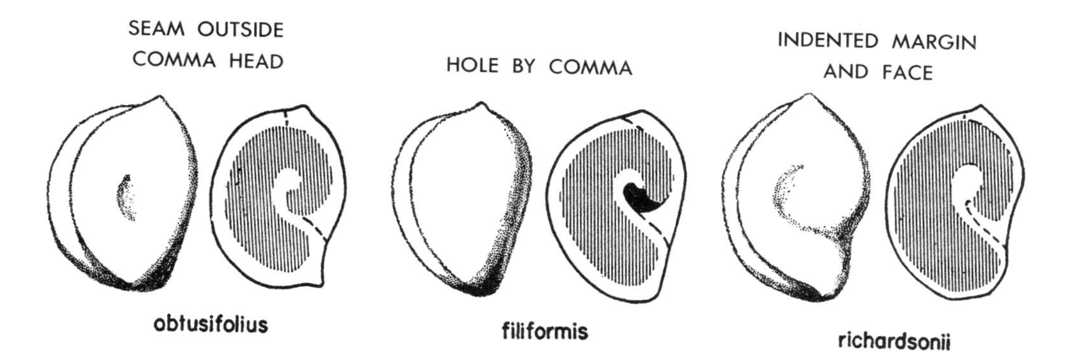

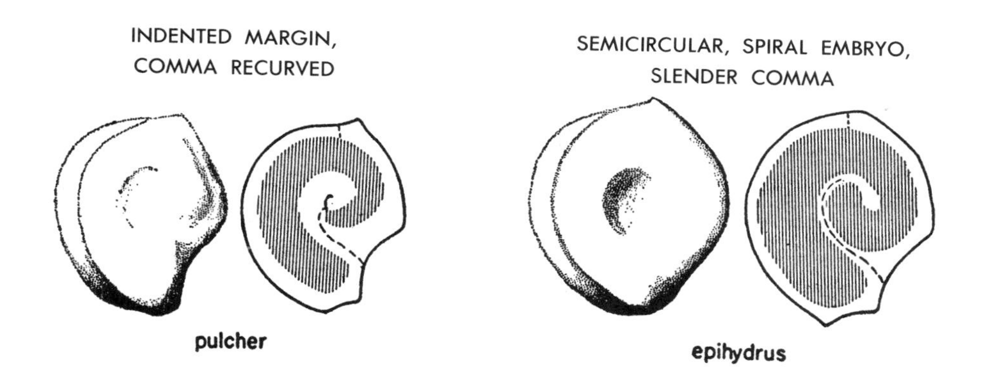

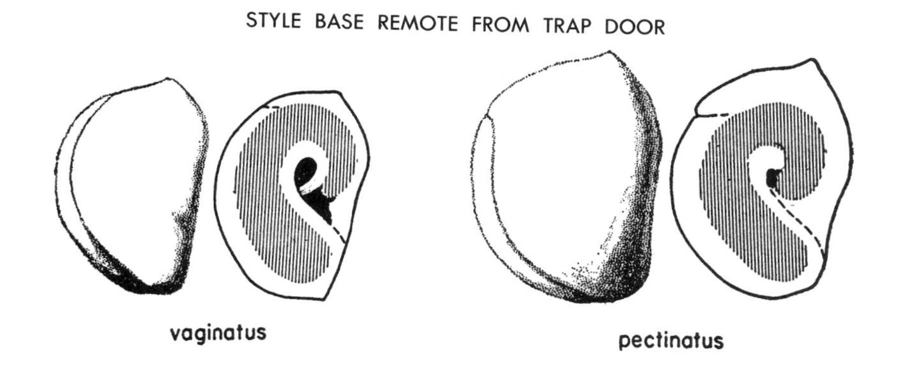

Fig. 15. Pondweed (*Potamogeton*) seeds

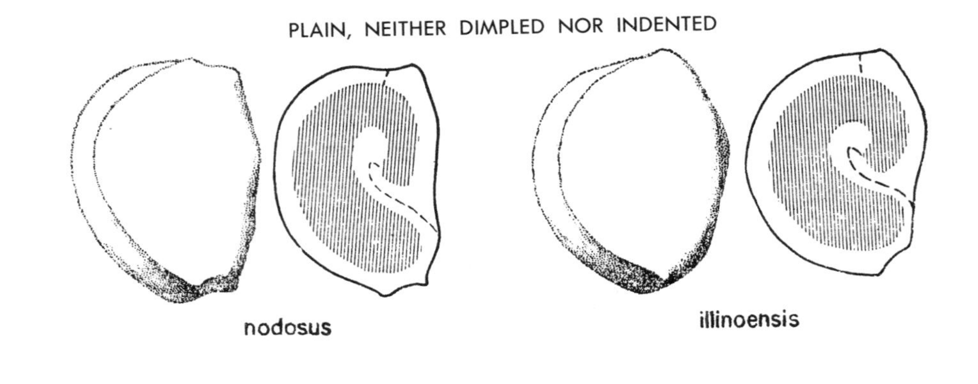

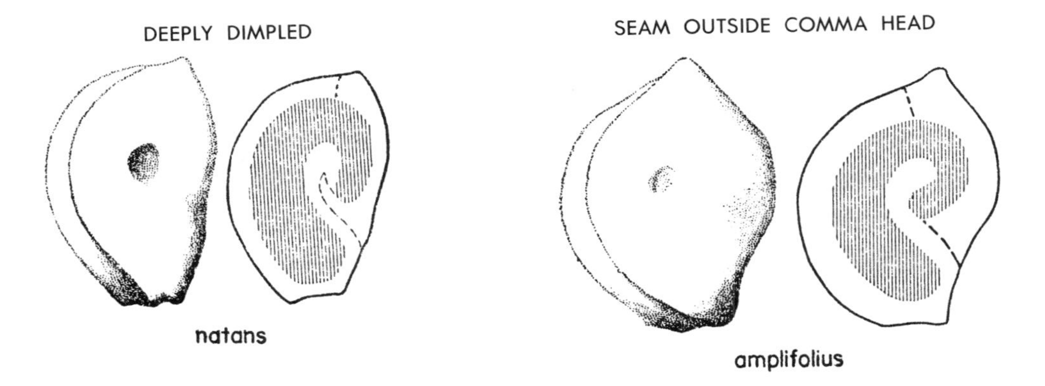

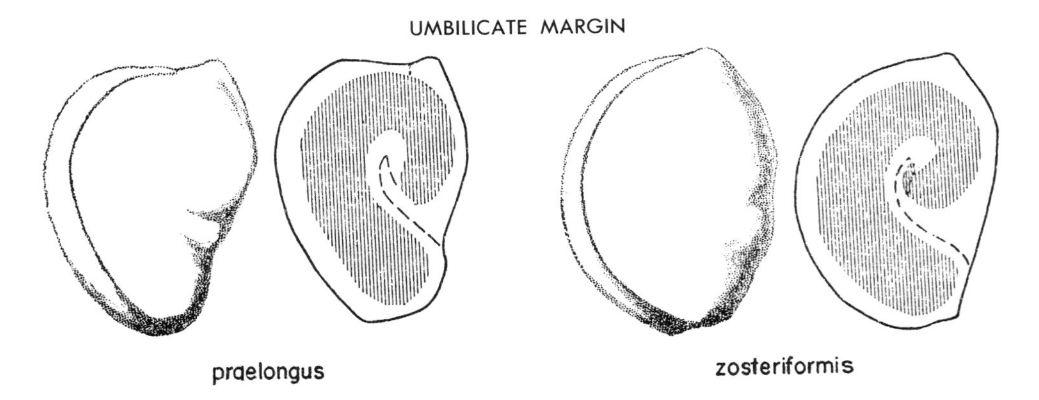

Fig. 16. Pondweed (*Potamogeton*) seeds

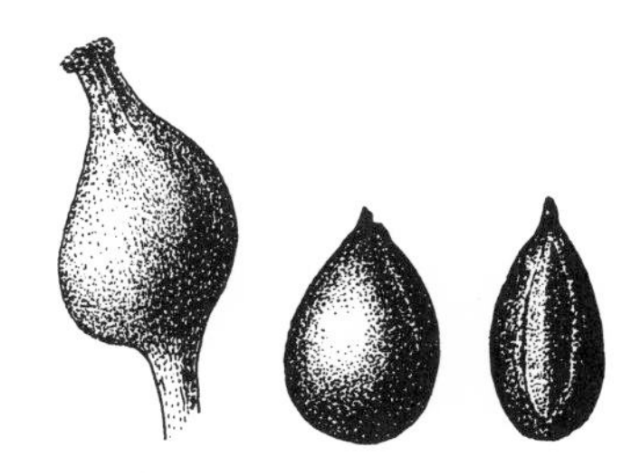

Fig. 17. *Ruppia maritima*

RUPPIA Widgeongrass (pl. 540)

The seeds are ovoid; minute (about 2 mm. long); and somewhat pointed at one end. When the dull, dark, skinlike covering is removed, the seeds are **black** and show a small **"trap door"** (fig. 17), similar to that of *Potamogeton* seeds.

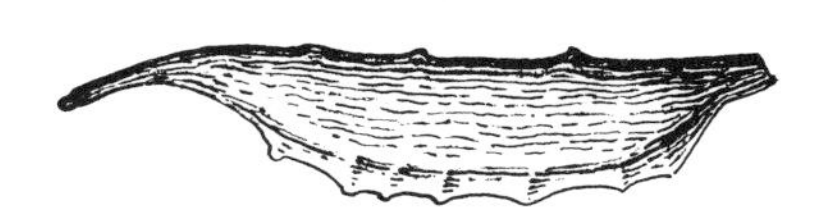

Fig. 18. *Zannichellia palustris*

ZANNICHELLIA Hornpondweed (pl. 541)

Seed body 2–3 mm. long, flattish, oblong-arched, with a stalk at one end and a style base, of variable length, at the other. Removal of the outer coating exposes a **series of spinelike projections** on both edges of the seed.

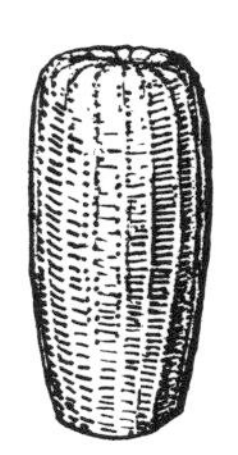

Fig. 19. *Zostera marina*

ZOSTERA Eelgrass (pl. 542)

Seeds oval-cylindric; about 3 mm. long; generally with about 20 distinct **lengthwise ribs** (fig. 19), though in some Pacific coast strains the ribs are obscure or absent.

NAJAS Naiad (pls. 544–546)

Seeds oval-oblong, somewhat pointed, reticulate-surfaced in common species, but smooth in the larger (4–5 mm.), darker, more localized species *N. marina*. The seeds of both *N. flexilis* and *N. guadalupensis* are about 3 mm. long, the latter proportionately broader and dull-surfaced, with **coarse reticulations,** whereas *flexilis* is smooth-surfaced and shiny, with **less conspicuous reticulations.**

ALISMACEAE Waterplantain Family

SAGITTARIA Arrowhead (pls. 623–625)

Species of *Sagittaria* can often be distinguished by characteristics of their thin, **flattish or winged** seeds; in particular by the **style beak's** location, direction, and length; by ridges on the surface; and by outline and dimensions of the **encircling wing** (figs. 20, 21). The embryo is linear U-shaped, with no endosperm.

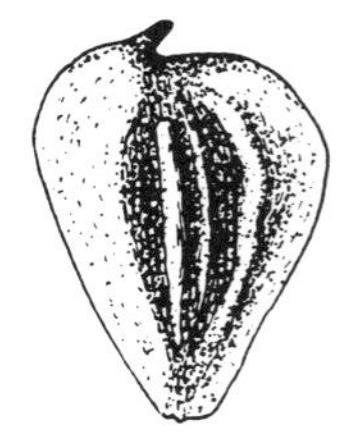

Fig. 20. *Sagittaria cuneata*

Fig. 21. *Sagittaria platyphylla*

HYDROCHARITACEAE Frogbit Family

Seeds small, thin-walled, and lacking endosperm.

ANACHARIS (ELODEA) Waterweed, Elodea (pl. 543)

Often the plants are sterile and bear none of the irregular fruits and **smooth, cylindric** seeds shown in figure 22; endosperm is lacking.

VALLISNERIA Wildcelery (pl. 547)

The delicate, **cellular-reticulate, undulate** coating (fig. 23, left) on the minute, rodlike seeds of wildcelery frequently show **iridescent** coloring under magnification. Faint longitudinal lines show when the outer coat is removed; endosperm none.

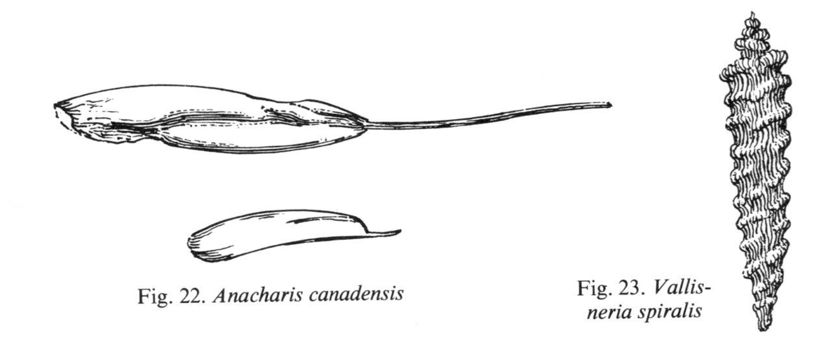

Fig. 22. *Anacharis canadensis*

Fig. 23. *Vallisneria spiralis*

GRAMINEAE Grass Family

Identification of grass seeds (grains or caryopses) is complicated by the large number of genera and species in the family, by considerable overlapping in seed characteristics of different tribes and genera, and by the fact that the seeds may be either clothed (enclosed in lemma and palea of the floret) or naked. An excellent government manual of the grasses of the United States[2] illustrates and describes seeds of more than 1,300 grass species. The comments here are, therefore, confined mostly to important clues for identification of the family and some of its genera and tribes. Characteristic seed features of various grass genera are shown in figure 24 and in seventy-three plates (exclusive of the duplicates in Weeds, Arranged by Physical Features), namely, plates 1–44, 511–520, 560–577, and 804.[3]

A distinctive feature of grass grains is the **lateral or basal-lateral embryo area.** The embryo area can be recognized by differences in color, as well as by furrows, ridges, or other irregularities which bound it. Lengthwise sections of seeds are useful in distinguishing genera. In many genera the embryo is confined to the lowest part of the seed, whereas in many others it extends about halfway up. In still others, such as *Bouteloua, Spartina,* and *Setaria,* it reaches all the way from the bottom to the top. Endosperm in grass seeds varies from floury to flinty or rarely fleshy, but in all cases it gives a starch reaction with iodine.

Especially useful identification pecularities of different grass seeds are generally evident at or near their point of attachment. In many grass genera there is a short, erect stalk (rachilla) at the base of the grain-enclosing floret—ordinarily on its concave or flattish side. This stalk is characteristic of the tribe

[2]Manual of the Grasses of the United States, by A. S. Hitchcock, 2d ed., rev. by Agnes Chase, U. S. Dept. Agr. Misc. Publ. 200, 1950.

[3]Forty-one genera of the Gramineae are represented in the photographic part of this manual. Because these genera are not described under individual headings, plate references to particular plants are not given here, as they are elsewhere throughout Identification Clues, but the reader can locate them through the Index. The generic and corresponding common names of the plants included are:

Agropyron (Quackgrass), *Agrostis* (Redtop), *Andropogon* (Broomsedge, Bluestem), *Aristida* (Three-awn), *Avena* (Oats, Wild Oats), *Bouteloua* (Grama), *Bromus* (Chess, Rescuegrass, Cheatgrass), *Cenchrus* (Sandbur), *Cynodon* (Bermudagrass), *Dactyloctenium* (Crowfootgrass), *Digitaria* (Crabgrass), *Distichlis* (Saltgrass), *Echinochloa* (Wildmillet), *Eleusine* (Goosegrass), *Eragrostis* (Stinkgrass), *Festuca* (Fescue), *Glyceria* (Mannagrass), *Holcus* Velvetgrass), *Hordeum* (Barley, Little Barley), *Leersia* (Rice Cutgrass), *Leptochloa* (Sprangletop), *Leptoloma* (Leptoloma), *Lolium* (Ryegrass), *Muhlenbergia* (Nimblewill), *Oryza* (Rice), *Panicum* (Panicum, Witchgrass, Proso, Wartgrass, Switchgrass, Browntop), *Paspalum* (Paspalum, Bullgrass, Knotgrass), *Phalaris* (Canarygrass), *Phleum* (Timothy), *Poa* (Annual Bluegrass, Kentucky Bluegrass), *Secale* (Rye), *Setaria* (Bristlegrass), *Sorghum* (Sorghum, Johnsongrass, Sudangrass), *Spartina* (Cordgrass), *Sporobolus* (Dropseed), *Stipa* (Needlegrass), *Tridens* (Purpletop), *Triticum* (Wheat), *Uniola* (Uniola), *Zizania* (Wildrice), and *Zizaniopsis* (Giant Cutgrass).

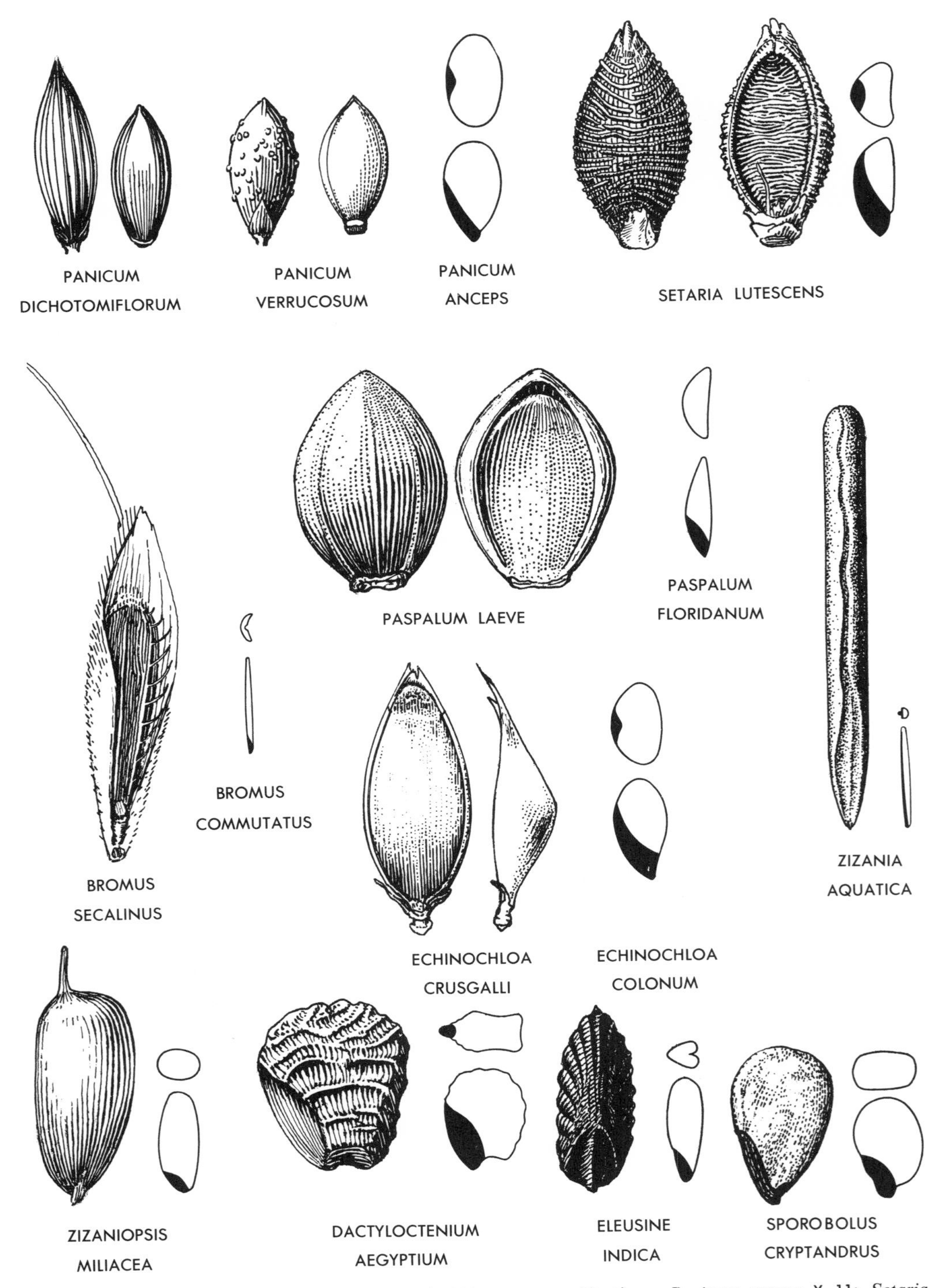

Fig. 24. Examples of various types of grass seeds. Diagram magnifications: *Panicum anceps* X 11; *Setaria lutescens* (right) X 7; *Paspalum floridanum* X 6; *Bromus commutatus* X 2; *Echinochloa colonum* X 9; *Zizania aquatica* (right) X 1.5; *Zizaniopsis miliacea* (right) X 6; *Dactyloctenium aegyptium* (right) X 14; *Eleusine indica* (right) X 13; *Sporobolus cryptandrus* (right) X 14.

Festuceae (represented in the photographic picture guide by *Bromus, Festuca, Glyceria, Poa, Eragrostis, Distichlis, Uniola,* and *Tridens*), but is also present in certain other tribes. Differences in width, length, shape, and pubescence of the stalk are often helpful for identification. Absence of the stalk is also significant.

Grooved grains are characteristic of certain genera in several tribes, including Hordeae, Agrostideae, and Aveneae. Variously awned lemmas also help identify seeds in a number of tribes. In most grasses, the lemma is somewhat boat-shaped, is rather loose, and extends beyond the grain. In the Paniceae, however, both the lemma and the palea are hard and enclose the grain tightly. Many of the Paniceae seeds are flattish on one side (plano-convex). This is particularly true of *Paspalum*. When outer coverings are removed from the seeds of *Panicum, Paspalum, Setaria, Digitaria,* and other Paniceae, a dark, roundish dot is evident near one end of the flattish face. In other tribes, certain genera, such as *Eragrostis, Sporobolus, Eleusine, Uniola,* and *Agrostis*, have seeds which can be freed easily from their coverings.

Seed-coat remnants of cultivated cereals consumed by animals can be recognized fairly readily. Corn seed-coat fragments are very persistent, are **semitransparent**, and show **faint straight lines**. Worn wheat coats generally have **fine hatchings or crosshatchings** on the thick reddish brown walls. Oats can be recognized by any remains of the **slender rachilla, bearded on both edges**; by the long, stiff hairs on the grain's surface; or by the narrow units into which the hulls break, usually with a narrow, nervelike line running lengthwise on the concave side.

CYPERACEAE Sedge Family

External features of Cyperaceae seeds (achenes) are diverse. Many of these seeds are 3-angled; many others are lens-shaped to globular; some have a cap (tubercle) on the top; some have their exteriors marked by minute papillae in rows; while still others are smooth or have various ridges, wrinklings, or other markings. Bristles are attached to the seed base in several genera. Of the genera considered here, *Dulichium, Eleocharis, Rhynchospora,* and *Scirpus* have **bristles,** but *Carex, Cladium, Fimbristylis,* and *Scleria* lack bristles.

Certain internal features, however, are fairly diagnostic. Most of the genera are similar in having tough walls with a **fibrous, finely striate inner surface.** Also, the **embryo is basal** (extending less than half the length of the seed) in **starchy endosperm,** which is granular to firm and semitransparent.

CYPERUS Flatsedge (pls. 45, 578–582)

Generally **triangular or elongate-triangular,** but the seeds of some species are lens-shaped; size varies from medium small to minute. In most species, the surface is **checkrowed diagonally with minute, papillae-like, cellular markings** (fig. 25).[4]

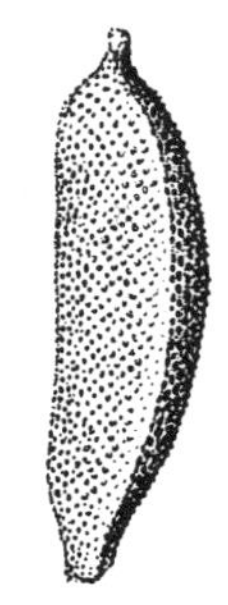

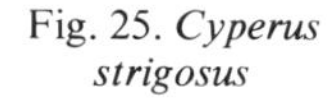

Fig. 25. *Cyperus strigosus*

Fig. 26. *Cyperus filiculmis* X 10

DULICHIUM Dulichium (pl. 583)

Linear-oblong, lens-shaped in cross section or obscurely angled; about 3 mm. long; the **long, stout style beak,** originally as long as the body, is persistent.

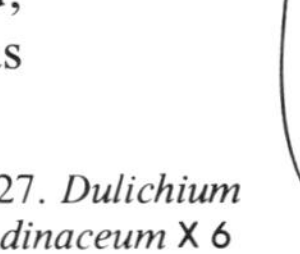

Fig. 27. *Dulichium arundinaceum* X 6

ELEOCHARIS Spikerush (pls. 584–594)

Seeds vary from triangular or ovoid to ovate-lens-shaped, and from medium small to minute. The **cap (tubercle)** is distinctive for this genus and *Rhynchospora.* The surface, in most of the species, has cellular reticulations. See figure 28.[5]

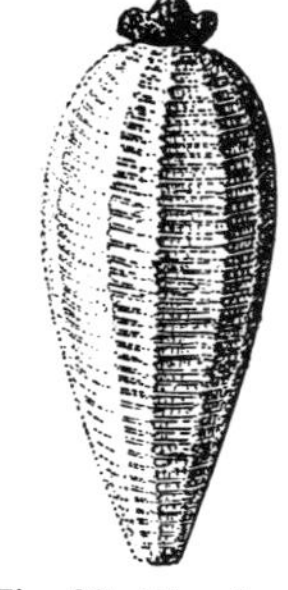

Fig. 28. *Eleocharis acicularis*

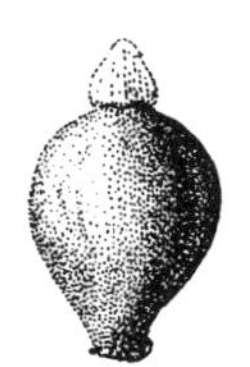

Fig. 29. *Eleocharis palustris*

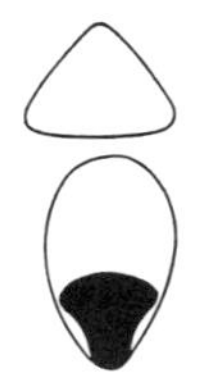

Fig. 30. *Eleocharis albida* X 22

FIMBRISTYLIS Fimbristylis (pls. 595, 596)

Seeds of this genus are diverse and lack distinctive group characteristics; they vary from ovate and lens-shaped to triangular, from medium small to minute, and from reticulate-surfaced to smooth.

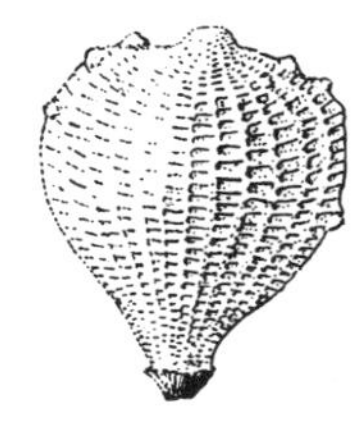

Fig. 31. *Fimbristylis baldwiniana*

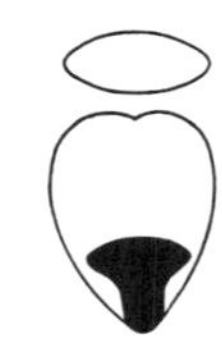

Fig. 32. *Fimbristylis castanea* X 13

---

[4]See also figures 124–135 in A Flora of the Marshes of California, by Herbert L. Mason, Univ. Calif. Press, 1957. This work is cited hereafter as Mason.

[5]See also the drawings in Mason, figs. 136–144.

RHYNCHOSPORA Beakrush (pls. 597, 598)

Like *Eleocharis*, this genus has a **cap (tubercle)** on top of its seeds. Generally the cap is pointed and beaklike, but on some species it is obtuse or rounded; surface varies from **smooth to transversely wrinkled;** size ranges from large to minute.

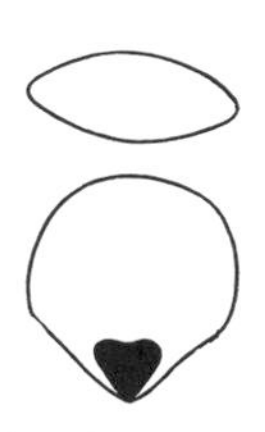

Fig. 33. *Rhynchospora cymosa* X 15

SCIRPUS Bulrush (pls. 599-612)

Seeds are generally ovate in outline and vary from **triangular or plano-convex** to lens-shaped, from smooth-surfaced to wrinkled in a few species, and from medium-sized to minute. The **blunt or pointed style base** helps distinguish bulrush seeds from seeds of other Cyperaceae. Variations in embryo shape are indicated in figures 37 and 38.

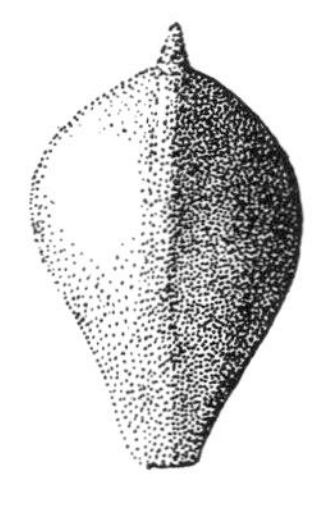

Fig. 34. *Scirpus americanus*

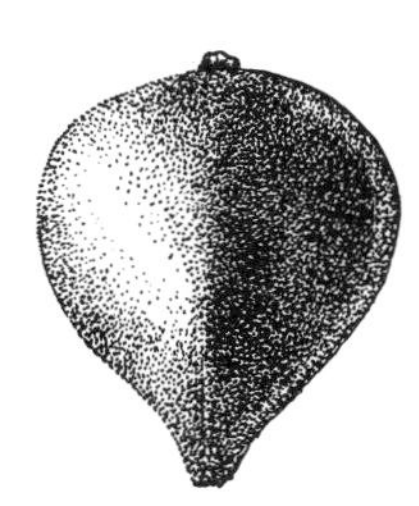

Fig. 35. *Scirpus robustus*

Fig. 36. *Scirpus paludosus*

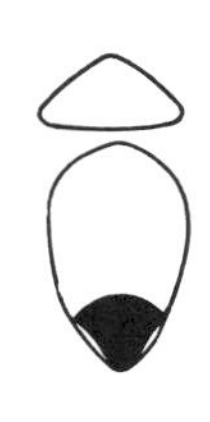

Fig. 37. *Scirpus atrovirens* X 25

Fig. 38. *Scirpus acutus* X 11

SCLERIA Scleria (pls. 613, 614)

Globular; with hard, **glassy or bonelike wall;** surface pitted-reticulate, wrinkled or smooth; size varies considerably in different species.

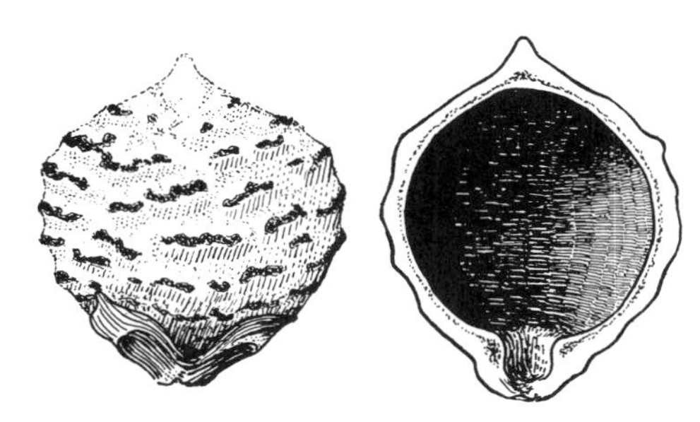

Fig. 39. *Scleria torreyana*

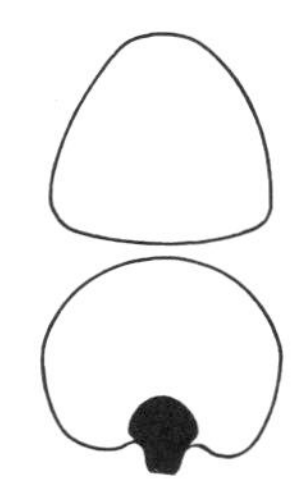

Fig. 40. *Scleria verticillata* X 22

CLADIUM Sawgrass (pl. 615)

Seeds **ovoid-ellipsoid; dark brown;** shallowly pitted on the surface. When the moderately soft outer coating is rubbed off, as by digestion in a duck stomach, a **hard, black ball** is left. Seeds of *C. jamaicense* and *mariscoides* are very similar.

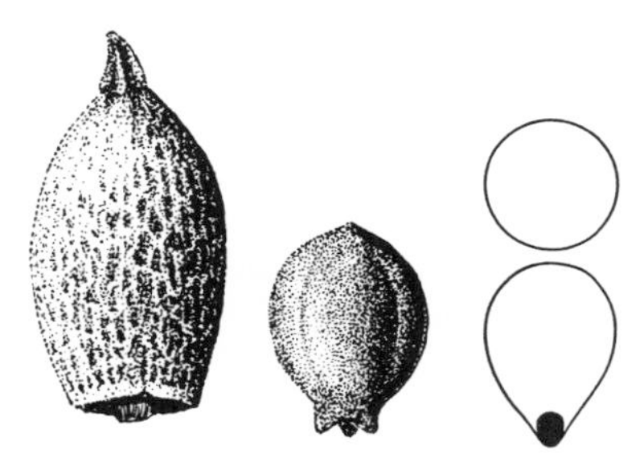

Fig. 41. *Cladium jamaicense.* Diagram (right) X 23

CAREX Sedge (pls. 616–619)

A unique feature of this genus is the **saclike structure (perigynium) covering its seeds** (fig. 42).[6] The perigynium usually has a **hollow, notched beak** of variable length, though in some species it is lacking or practically so. Lengthwise ribs may be prominent or absent, and the perigynium may either envelope the seed loosely or fit so tightly as to appear part of it.

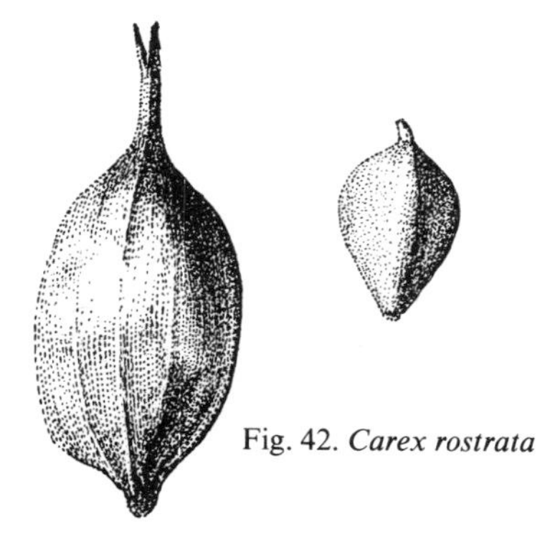

Fig. 42. *Carex rostrata*

Carex seeds vary from triangular to lens-shaped and from small to large. Their comparatively smooth surface distinguishes them from those species of *Cyperus* which have checkrowed markings; but small, light-colored, elongate-triangular *Carex* seeds are sometimes difficult to tell from similar seeds of a few species of bulrush. If bristles are present, they rule out *Carex*.

Fig. 43. *Carex lurida* X 9

COMMELINACEAE Dayflower Family

Seeds flattish to ovoid or angular, often truncate at one end and rounded at the other; embryo small and in a special chamber, usually outlined on the surface by a **circular, depressed area** with a central nipple; hilum a distinct **narrow line** on a flattish side; endosperm starchy.

---

[6]Also see the drawings in Mason, figs. 86–123.

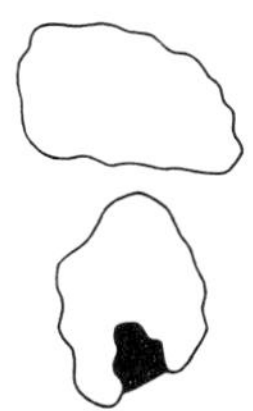

ANEILEMA Aneilema (pl. 626)

The seeds are **like small editions of Commelina seeds**; 1.5–4 mm. long; flattish-elliptic with one end truncate; embryo area embedded in an edge; hilum line conspicuous on the flattest side.

Fig. 44. *Aneilema nudiflorum* X 14

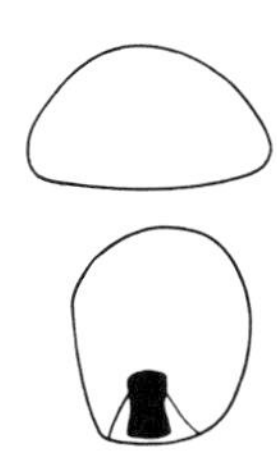

COMMELINA Dayflower (pls. 46, 47)

Shape variable, from flattish and elliptic-truncate to ovoid or nearly globular. *C. crispa* (fig. 45) represents the semiglobular type with a plane or flattish surface bearing the hilum line. Its embryo area is obscure and not sunken. *C. virginica* is flattish, large, up to 5 mm. long, with its embryo area buried on an edge. Its hilum line is evident on the flatter face.

Fig. 45. *Commelina crispa* X 8

TRADESCANTIA Spiderwort (pl. 48)

Two surface features distinguish this genus from *Commelina* and *Aneilema*: (1) its **embryo area is deeply sunken** into a broad face, which is (2) **sculptured with radial notches**. The embryo is also distinct in being capitate (fig. 46).

Fig. 46. *Tradescantia virginiana* X 13

PONTEDERIACEAE Pickerelweed Family

Seeds **elliptic-cylindric;** obscurely cellular-sculptured; **embryo linear-axial in starchy endosperm.**

EICHHORNIA Waterhyacinth (pl. 548)

Minute, about 1 mm. long; oblong-cylindric with **one end truncate** and the other somewhat pointed; surface cellular-reticulate rather irregularly.

PONTEDERIA Pickerelweed (pl. 627)

Fig. 47. *Pontederia cordata* X 5

Ovoid-ellipsoid; about 3–4 mm. long; with **20–25** obscure, slightly raised **lengthwise lines** and a shallow, **crater-like depression** at each end. The dry, seedlike fruit has several notched wings or ridges running lengthwise and is tipped by an arched style beak.

JUNCACEAE Rush Family

JUNCUS Rush (pls. 628, 629)

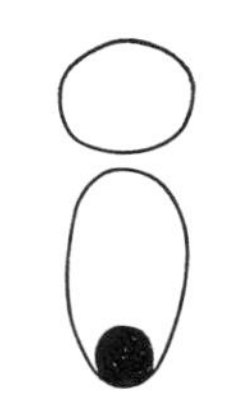

Fig. 48. *Juncus marginatus* X 43

Very minute, generally less than 0.5 mm. long; **ellipsoid to globular;** commonly marked with **cellular reticulations** and sometimes bearing terminal appendages or low, winglike ridges.[7] The embryo is basal in starchy endosperm.

LILIACEAE Lily Family

External features of seeds in this large family are too diverse to characterize readily. The seeds vary from globular to flat and winged, or irregular, and also range considerably in size. Internally, they are uniform in having a **central-axile embryo** (generally linear, but short or basal in some species) in fleshy or hard endosperm.

SMILACINA False Solomonseal (pl. 805)
POLYGONATUM Solomonseal (pl. 806)
MAIANTHEMUM Beadruby (pl. 807)

These three genera are similar in having **subglobose** or **globose** seeds which are often slightly translucent. Usually they have a darker circular area, and, at the opposite pole, an obscure, nipple-like elevation. Endosperm is hard.

Smilacina seeds are largest, averaging 3–4 mm. across. The embryo is **more than half the seed length** in several species (fig. 49), but less than half in *S. racemosa.*

---

[7]See drawings in Mason, figs. 169–184.

Fig. 49. *Smilacina stellata* X 5

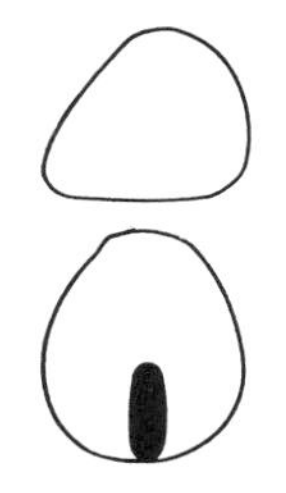

Fig. 50. *Polygonatum commutatum* X 5

Fig. 51. *Maianthemum canadense* X 5.5

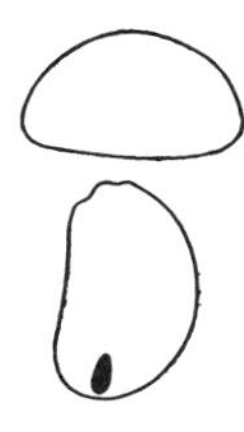

Fig. 52. *Smilax glauca* X 4

Polygonatum is similar but smaller, averaging 2–3 mm. across, and its embryo is **less than half the seed length** (fig. 50) in all species.

Maianthemum is comparatively small, usually about 2 mm. across, and its embryo is **at least half the length of the seed** (fig. 51).

SMILAX Greenbrier (pls. 684, 685)

Seeds subglobose, hemispheric, or broadly rounded-triangular; 4–5 mm. in longest dimension; usually dark reddish brown, with a **circular dark area;** embryo is small, basal-linear, in **very hard, semitranslucent endosperm.**

AMARYLLIDACEAE Amaryllis Family

HYPOXIS Yellow Stargrass (pl. 808)

Subglobose; 1 mm. long or slightly longer; **black,** shiny; covered closely with **papillae or toothlike projections** and commonly with a **handle-like structure**. Wall brittle; endosperm fleshy.

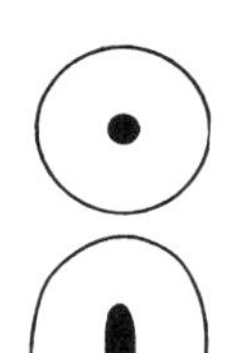

Fig. 53. *Hypoxis hirsuta* X 15

IRIDACEAE Iris Family

SISYRINCHIUM Blue-eyed-grass (pl. 630)

Small, 0.5–1 mm. across in most species, 1–1.5 mm. in *S. bellum;* **globular to rounded-angular; black** or dark, surface varies from distinctly **pitted or reticulate to smooth**, or practically so; the attachment zone is a puckered-up depression; endosperm hard and semi-transparent or whitish.

Fig. 54. *Sisyrinchium bellum* X 10

MYRICACEAE Waxmyrtle Family

MYRICA Waxmyrtle, Bayberry (pl. 686)

Seeds (drupes) globular or ovoid, coated by a thick layer of **whitish, waxy material;** diameter of two common species is: 2–3 mm. in *M. cerifera* and 3–4 mm. in *M. pensylvanica.* Distinctive of the drupe's thick, woody wall, as viewed in section, is the comparatively thin zone at each end (fig. 55).

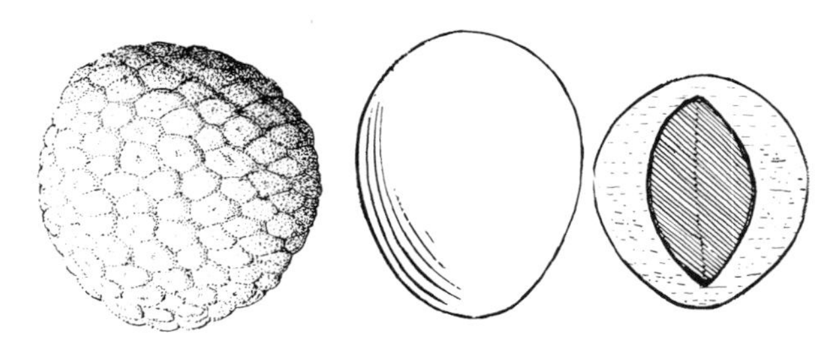

Fig. 55. *Myrica cerifera*

BETULACEAE Birch Family

A consistent family characteristic is the presence of a small amount of endosperm (more translucent than the embryo) in the pointed, basal end of the seeds (nutlets or nuts). Even *Corylus* has a little. Most seeds of *Betula* and some of *Alnus* are encircled by a thin wing.

OSTRYA Hop-hornbeam (pl. 688)

Seeds compressed lanceolate-ellipsoid, long-pointed; averaging 4 X 2.2 X 1 mm. This genus, as well as *Carpinus* and genera of the pine and beech families, have a **thin, brownish, papery coat** (tegmen) lining the inside of the woody wall. The tapering shape and smooth surface of *Ostrya* distinguish it from *Carpinus*.

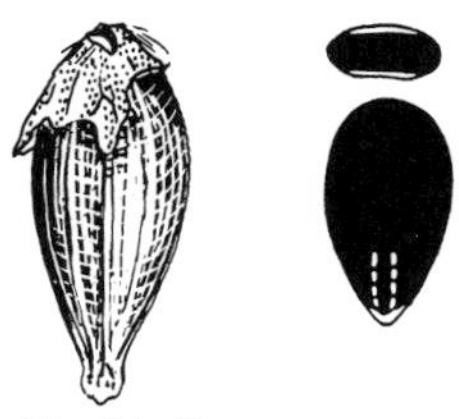

Fig. 56. *Ostrya virginiana.* Diagram (right) X 4

CARPINUS Hornbeam (pl. 687)

Resembles *Ostrya* in the **brownish, papery lining** inside the wall, but differs in shape (ovate-lenticular), size (averaging 2.7 X 2.4 X 1.2 mm.), and the **3 or 4** prominent **longitudinal ridges on both faces,** in addition to ridges along the margins.

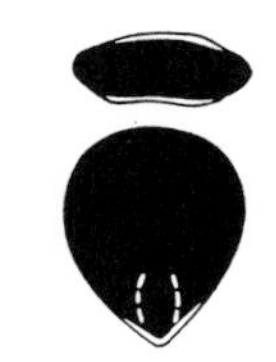

Fig. 57. *Carpinus caroliniana* X 5.5

BETULA Birch (pl. 689)

The nearly **encircling wing and 2 persistent stigmas** help identify birch seeds. Different species can be distinguished by characters such as outline of the seed body (oval, ovate, or obovate) and breadth of the wings.[8]

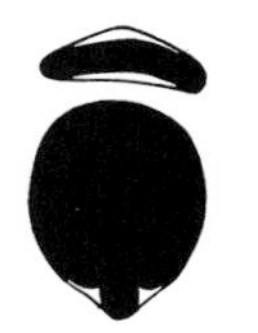

Fig. 58. *Betula lenta* X 9

ALNUS Alder (pl. 690)

Flat, thin, less than 1 mm. thick. In contrast to *Betula*, alder stigmas tend to break off and leave **2 stubby style bases**. Winged species such as *A. rubra* and *sinuata* resemble birch seeds; wingless alders such as *incana* and *serrulata* have thicker, woody, flat faces.

Fig. 59. *Alnus serrulata* X 7.5

FAGACEAE Beech Family

The woody-walled nuts of this family are borne in distinctive **cups** or husks. A papery layer, varying from smooth to woolly, lines the inside of the wall; endosperm is absent, the nut interior being almost completely occupied by the 2 thick cotyledons.

---

[8] As indicated, for example, in A Seed Key for Five Northeastern Birches, by Frank E. Cunningham, Jour. For. 55:844–845, 1957.

FAGUS Beech (pl. 691)

Beechnuts are **sharply triangular** and have **folded cotyledons**. The reddish brown, papery lining inside the wall tends to adhere to the cotyledons.

Fig. 60. Beechnuts and husk

QUERCUS Oak (pls. 692–696)

Range in size and shape of oak nuts (acorns) is indicated in figure 61. Cups at the base of the acorns are diverse too. They may cover much of the acorn, as in *Q. lyrata* and *macrocarpa*, or may be very shallow, as in *palustris* and *borealis*. In some species, the cup scales are small and tightly imbricated, whereas in others the scales are large and loose. Additional features which help identify different oaks by their acorns are peculiarities in the large basal scar area, smoothness or roughness of the surface, the style base and markings near it, and the amount (or absence) of **hairy matting** inside the brownish membrane lining the interior. Acorns of many oak species are illustrated in the special work on trees published as the Department of Agriculture's yearbook for 1949,[9] as well as in major floras and in other tree books.

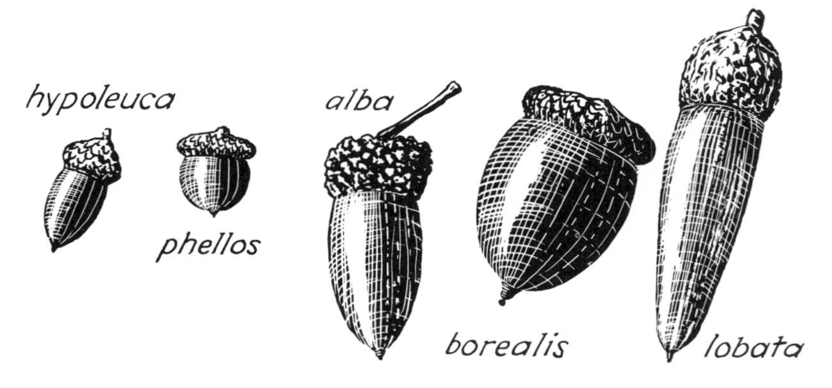

Fig. 61. Representative acorns

[9]Trees: The Yearbook of Agriculture, 1949, published by the U. S. Dept. Agr., 1949.

ULMACEAE Elm Family

The three native genera, *Ulmus, Planera,* and *Celtis*, have seeds which are too diverse to identify by a generalized description. They vary from winged seeds (samaras) to drupes or nutlets. Endosperm is absent in some and present in others.

ULMUS Elm (pls. 697, 698)

Seeds flattish; **enclosed and encircled by thin wings,** except at apical slot, which is flanked by **incurved points**. The seed body is obovate, flat, and covered by a thin, brownish coat; endosperm is lacking, the embryonic axis being almost completely hidden in the 2 large cotyledons.

Fig. 62. Elm seeds

CELTIS Hackberry (pls. 699, 700)

Seeds (nutlets of drupes) **globose**; 3–7 mm. in diameter; **walls bony or somewhat glassy,** thick (about 0.5 mm.), and brittle; broken edges whitish or cream-colored; surface smooth or pitted, in different species. The wall interior has a yellowish brown coating which, under magnification, shows finely reticulate markings. The embryo is folded (fig. 63) and embedded in a small amount of semi-transparent endosperm.

Fig. 63. *Celtis reticulata* X 4

MORACEAE Mulberry Family

The only seed (achene) character which holds for the whole family is the arched or bent position of the embryo in fleshy endosperm.

MORUS Mulberry (pls. 702, 703)

Seeds oval-lenticular to ovoid; 2–3 mm. long; with a **light-colored, slender attachment stalk** appressed to the thinner edge; wall woody but thin; inner surface marked with cellular reticulations.

Fig. 64. *Morus alba* X 7.5

MACLURA Osage-orange (pl. 701)

Oblong-elliptic in outline and lens-shaped or flattish in section; **oblique-truncate** at one end, pointed at the other; a **narrow, low ridge borders the margins;** about 8 X 4.5 X 1.5 mm.; wall thin, interior somewhat glossy and not reticulate; endosperm scant.

Fig. 65. *Maclura pomifera* X 2

URTICACEAE Nettle Family

Seeds (achenes) rather small, generally 1–2 mm. long; diverse in shape and surface in different species, but similar in having a **spatulate embryo and scant endosperm.**

URTICA Nettle (pl. 809)

Barely 1 mm. long (averaging about 0.9 X 0.7 X 0.3 mm.); oval-ovate in outline, pointed at both ends, and compressed-lenticular in cross section; the enveloping 2- and 5-parted calyx often persists.

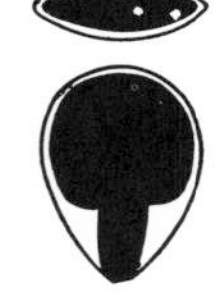

Fig. 66. *Urtica lyallii* X 17

BOEHMERIA Falsenettle (pl. 810)

Minute, about 0.7 X 0.5 X 0.3 mm.; ovate-compressed, with pointed style base; body with **horse-collar ring** of spongy tissue except at base; stubby, whitish hairs along edges or elsewhere.

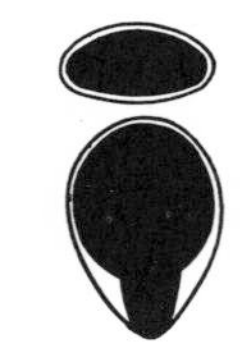

Fig. 67. *Boehmeria cylindrica* X 22

PARIETARIA Pellitory (pl. 811)

Small, averaging 1 X 0.6 X 0.5 mm.; oval-ovoid, slightly compressed; smooth and **shiny, like an insect egg; attachment scar whitish,** circular; other end pointed.

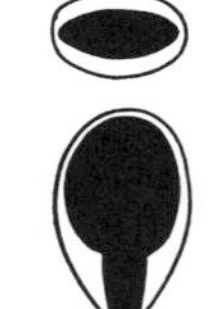

Fig. 68. *Parietaria pensylvanica* X 15

POLYGONACEAE Buckwheat Family

Seeds (achenes) of various sizes and shapes (many triangular or lens-shaped), but similar in the (usually) **peripheral, arched embryo** and **hard, semitransparent, starchy endosperm.** Outer wall (pericarp) firm, brittle; inner coat (testa) free from pericarp and adherent to the true seed.

ERIOGONUM Eriogonum (pls. 49, 50)

Generally small, about 1–2 mm. long; varies in shape from **decanter-like**, with a bulbous base and long neck, or **elongate-triangular**, to **ellipsoid-lenticular**; surface usually smooth and shiny, but a few, such as *E. alatum*, are rough and dull; some have pubescence toward apex; cross sections show a **sandwich-like arrangement** of embryo and endosperm.

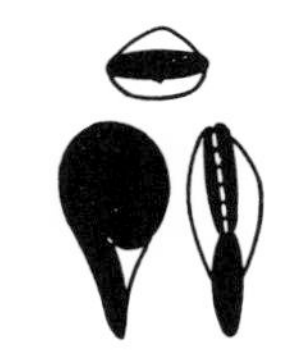

Fig. 69. *Eriogonum latifolium* X 7.5

RUMEX Dock (pls. 51, 52)

Usually **sharply triangular** (*R. acetosella* is rounded-triangular) and **acute-pointed** at apex; surface smooth, brownish, and generally shiny; wall thin. Cross sections show **embryo on middle of one side**, rather than in a corner as in *Polygonum. R. verticillatus* and *maritimus* are exceptions in not having a truly parietal embryo and in having cotyledons at right angles to the wall, instead of parallel to it.

Fig. 70. *Rumex crispus* X 9

POLYGONUM (inclusive sense) Smartweed, Knotweed (pls. 53–55, 631–642)

Mainly triangular, but some ovate in outline and flattish or lens-shaped; black, dark brown, or lighter brown; length varies from 2 to 5 mm.; surface smooth and shiny in many, but rough and dull in others; cotyledons vary from narrow to broad (figs. 71–73). Twenty-two species of *Polygonum* are illustrated on millimeter grids in figures 74–76.[10]

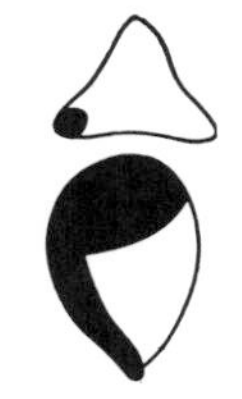

Fig. 71. *Polygonum punctatum* X 9

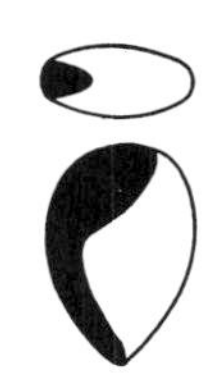

Fig. 72. *Polygonum virginianum* X 5

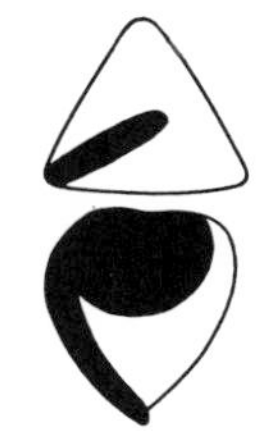

Fig. 73. *Polygonum sagittatum* X 7.5

[10]Reproduced from Identifying Polygonum Seeds, by Alexander C. Martin, Jour. Wildl. Mgmt. 18:514–520, 1954.

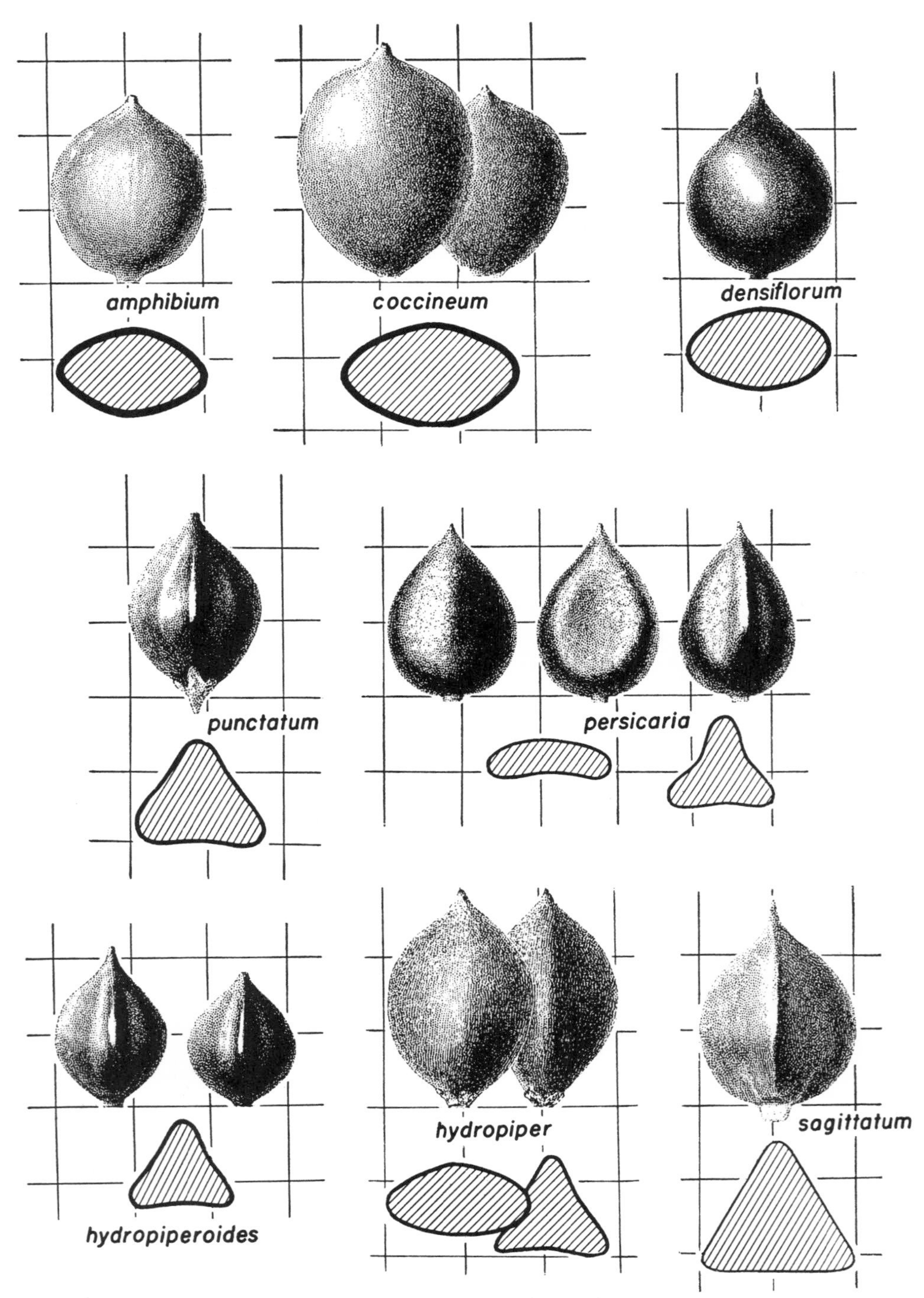

Fig. 74. Smartweeds (*Polygonum*)

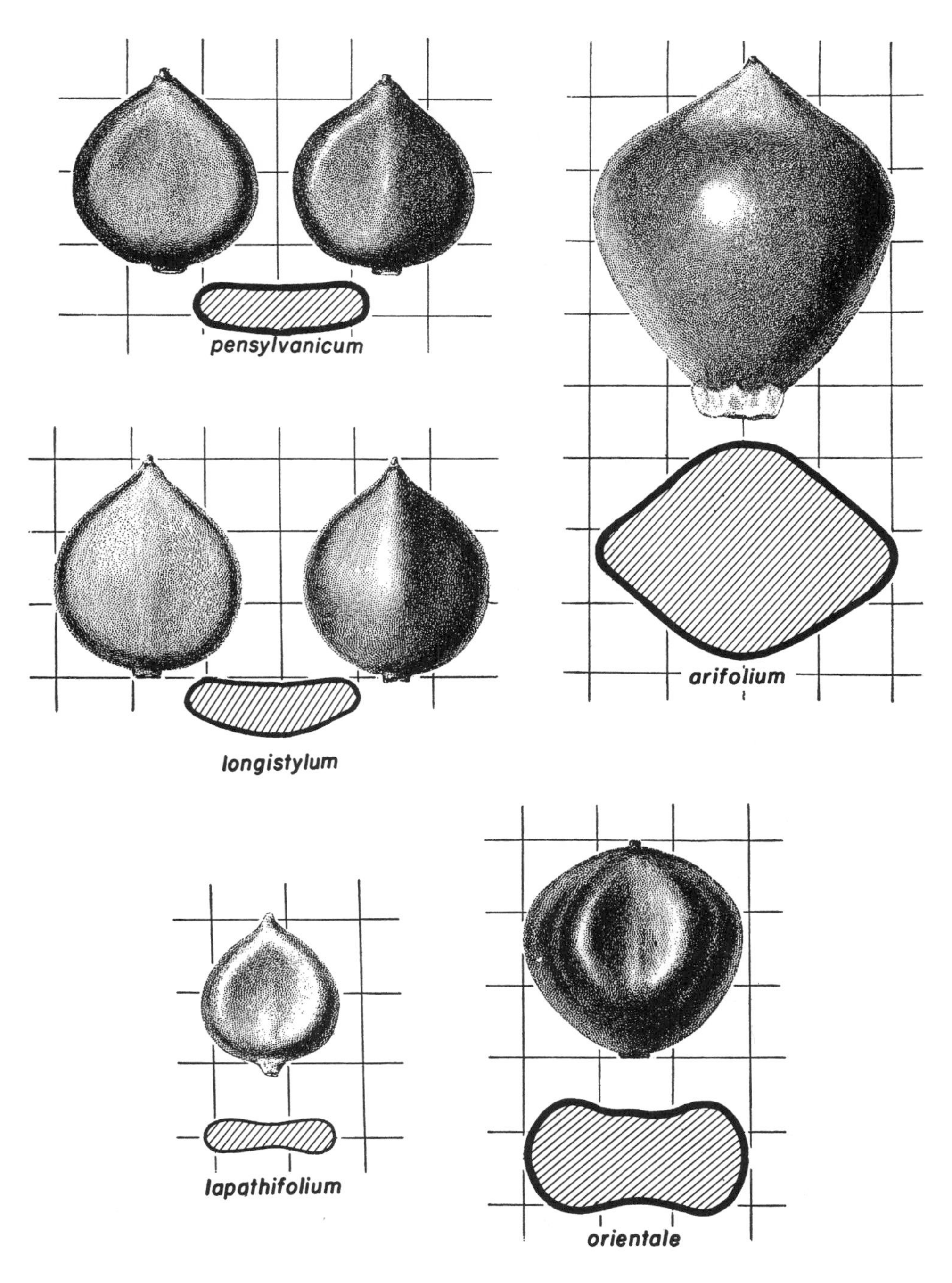

Fig. 75. Smartweeds (*Polygonum*)

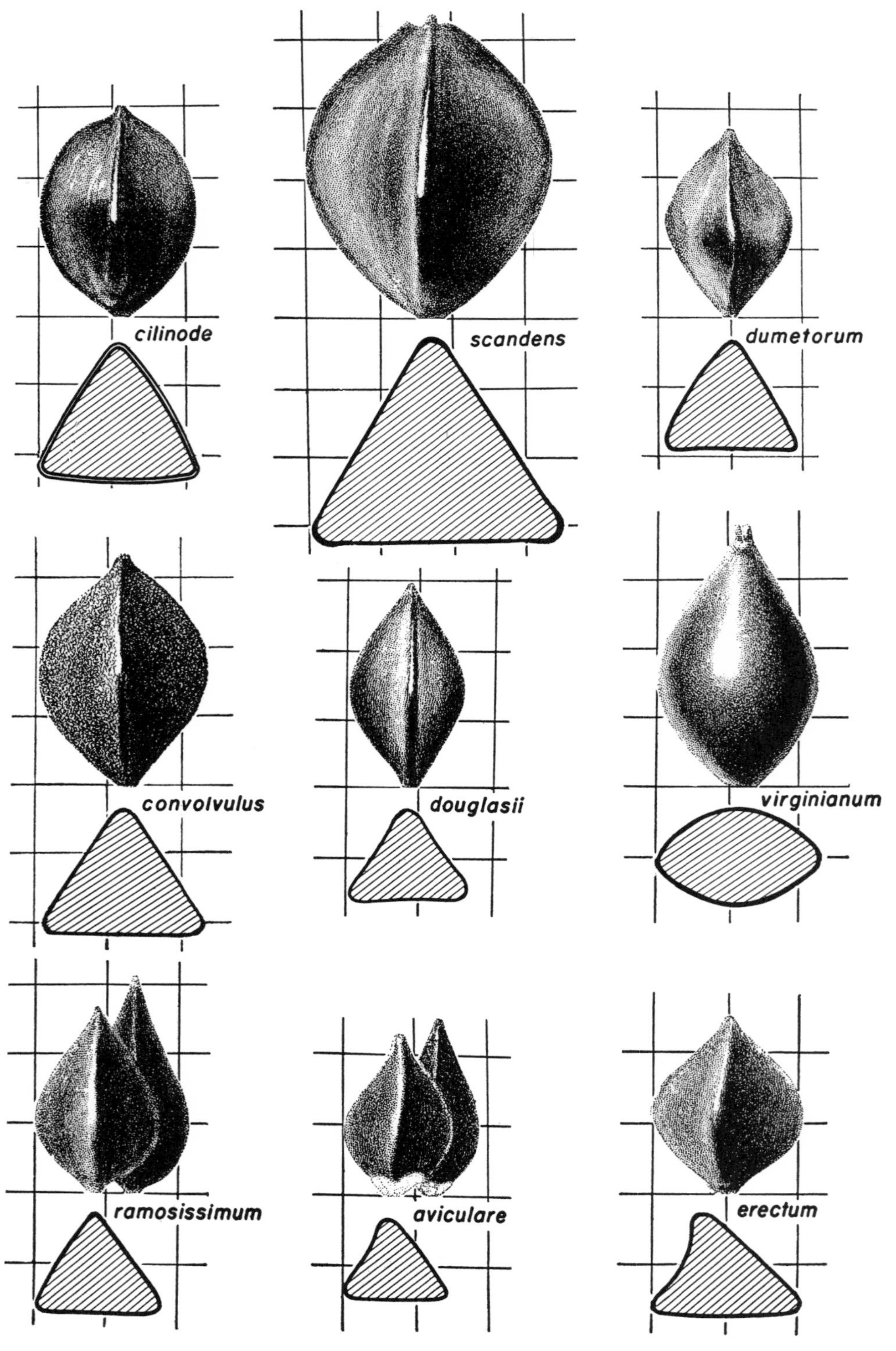

Fig. 76. Knotweeds (*Polygonum*)

FAGOPYRUM Buckwheat (pl. 521)

Seeds **sharp-edged triangular**; often marked with **zebra-like streaks**; average size 4.5 X 3.5 X 3.5 mm.; wall thin; the embryo is distinct in having its slender **cotyledons folded** irregularly through the endosperm.

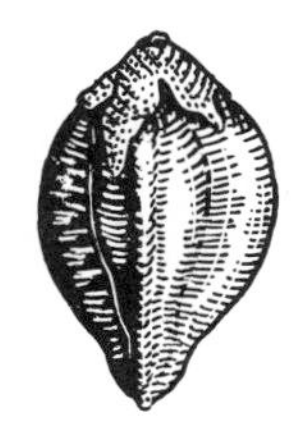

Fig. 77. *Fagopyrum esculentum.* Exterior view (left) X 5; diagram (right) X 3.5

CHENOPODIACEAE Goosefoot Family

Seeds (utricles) mainly **circular and compressed-lenticular**, but diverse in other external features. Surface usually gives some indication that **embryo is arched, curled into a ring, or spirally coiled**. Endosperm (starchy) present, except in *Salicornia, Salsola,* and *Sarcobatus.*

CHENOPODIUM Goosefoot (pls. 56–58)

Circular-lenticular; 1–2 mm. across; frequently with the **filmy calyx persisting as a cover**, otherwise rather shiny and black or dark brown. Generally distinct from *Amaranthus* in having a **rounded margin**, instead of a narrow rim. A notch or groove at one point on the margin varies from evident to obscure. Differences in embryo size and position are indicated in figures 78–80.

Fig. 78. *Chenopodium capitatum* X 17

Fig. 79. *Chenopodium ambrosioides* X 22

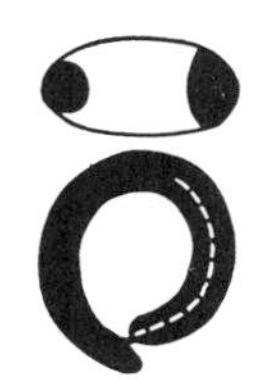

Fig. 80. *Chenopodium album* X 12

CYCLOLOMA Ringwing (pl. 59)

Seed enclosed in a flattish, circular, toothed, brown to purple **wing** (calyx), which is 3–5 mm. across. The concave side has 5 slot-like openings between rounded points, which spread into **the form of a star** as the seed matures. With the wing removed, the true seeds are circular-lenticular, rounded on margins, dull blackish, about 2 mm. across and 0.7 mm. thick.

Fig. 81. *Cycloloma atriplicifolium* X 12

ATRIPLEX Saltbush (pls. 60, 61)

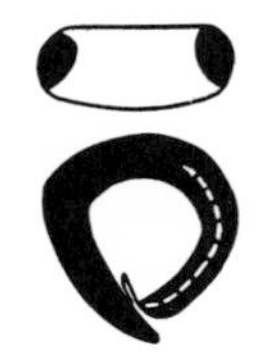
Fig. 82. *Atriplex rosea* X 9

Seeds enclosed commonly in a **pair of appressed bracts**, which vary greatly in size, shape, and marginal teeth in different species. The true seeds, removed from the bracts, are generally circular to oval, and lens-shaped or flat, and range in width from 1 to 3.5 mm. The outline of the **ringlike embryo** is evident on the exterior, along the margin, its radicle tip protruding to a different extent in different species. Good drawings of the seeds of many species of *Atriplex* are presented in Abrams' flora of the Pacific states.[11]

CORISPERMUM Tickseed (pl. 62)

Fig. 83. *Corispermum hyssopifolium* X 7

Oval in outline, flattish and slightly concave-convex, with a persistent **encircling wing**; surface glossy, as if varnished, splotched with rusty brown over seed-body area and lighter-colored along the marginal zone; about **4 X 3 X 0.5** mm. Base of wing emarginate or rounded, and apical end bearing **2 minute, toothlike style bases.**

KOCHIA Summer-cypress (pl. 63)

Fig. 84. *Kochia scoparia* X 9

Seeds **flat**, narrowly obovate, with a low, **wall-like ridge bordering the margin** on both faces, except at the narrower, basal end; approximately 2 X 1.5 X 0.5 mm.; coated by a delicate, grayish covering with 4 or 5 starlike points along the margin and bearing an indistinct nipple or puckered-up area in the middle of one face. **Embryo greenish** at apical end, as in *Eurotia*.

SUAEDA Seablite (pl. 643)

Fig. 85. *Suaeda depressa* X 14

Seeds lens-shaped, round-margined, and nearly circular except for a **slightly hooked projection and notch** on the margin; small, about 1 X 1 X 0.5 mm.; black or dark reddish brown; **glossy and finely cellular-reticulate.** The firm, semitransparent endosperm is limited to a zone on each side of the **spirally coiled embryo.**

SALSOLA Russianthistle (pl. 64)

Eastern variety narrowly, if at all, winged, whereas the common Western form is **broadly winged**, with thin, lacy extensions of calyx making the seed **top-shaped**; when calyx covering is removed, the seed consists solely of the **coiled, globose-obconic embryo** (no endosperm), about 2 mm. across.

---

[11]An Illustrated Flora of the Pacific States, by Leroy Abrams, Vol. 2, Stanford Univ. Press, 1944.

AMARANTHACEAE Pigweed Family

AMARANTHUS Pigweed, Waterhemp (pls. 65–67, 644, 645)

Mainly circular-lenticular, with its **edge narrowed into a thin rim** (*Chenopodium* is different in having a rounded edge); notch on margin often obscure; usually **shiny black** or dark reddish brown; most species are 1–1.5 mm. across, except the ovate *A. cannabinus*, which is about 3.5 mm. long and is flat and dull-surfaced; embryo curved around firm, starchy endosperm.

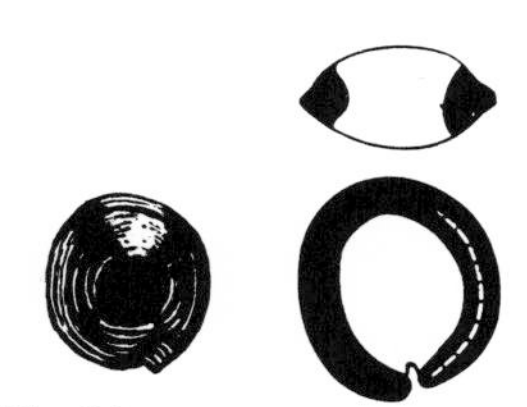

Fig. 86. *Amaranthus retroflexus*. Exterior (left) X 10; diagram (right) X 14

NYCTAGINACEAE Four-o-clock Family

Seeds (anthocarps) diverse externally; **embryo peripheral and arched** around the starchy endosperm, except in *Pisonia*, in which the embryo is not arched.

BOERHAAVIA Spiderling (pl. 68)

Shaped **like a punching bag**, obovoid to oblanceolate; usually 4- or 5-ridged and, in most species, with **appressed, whitish hairs** oriented lengthwise in grooves between ridges; yellowish to brown; 2–5 mm. long.

Fig. 87. *Boerhaavia torreyana* X 9

ABRONIA Sandverbena (pl. 69)

Ellipsoid to oblong; 2–6 mm. long; dark and shiny or brown and dull; enclosed in a **papery or woody covering with broad lobes or wide, thin wings**. A distinctive internal feature is its possession of **just one normal cotyledon** and **a rudimentary stub** of the other.

Fig. 88. *Abronia fragrans* X 6.5

OXYBAPHUS Umbrellawort (pl. 70)

Body **warty**; narrowly obovoid, with about **5 longitudinal angles or ridges**, near the base of which are whitish, appressed hairs oriented lengthwise. Most species also have a coating of fine woolly hairs. Olive-green to brown; 3–5 mm. long.

Fig. 89. *Oxybaphus albidus* X 4.5

PHYTOLACCACEAE Pokeberry Family

PHYTOLACCA Pokeberry (pl. 71)

This is the **largest black, shiny, circular-lenticular seed**; 2.8 X 2.5 X 1 mm. In effect, it is a much larger edition of *Amaranthus* or *Chenopodium*, with a similar marginal notch and a peripheral embryo encircling a hard, starchy endosperm. The **V-shaped marginal notch commonly has a whitish plug in it.**

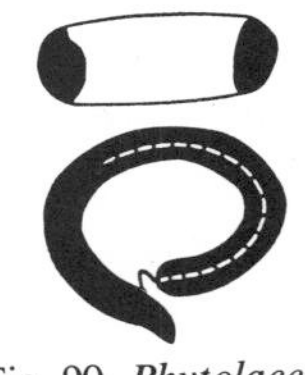

Fig. 90. *Phytolacca americana* X 6

AIZOACEAE Carpetweed Family

Seeds **minute or small**, generally lenticular or reniform; embryo curved peripherally around endosperm.

MOLLUGO Carpetweed (pl. 72)

Minute, 0.7 X 0.6 X 0.4 mm. lenticular-reniform, with a distinct marginal notch and **3–5 concentric ridges on and along the margin,** except in the notch area; surface shiny brown, with fine cellular reticulations.

Fig. 91. *Mollugo verticillata* X 19

SESUVIUM Sea-purslane (pl. 646)

Minute, 0.8 X 0.7 X 0.4 mm.; lenticular-reniform, with a marginal notch and adjoining projection; surface **dull black or dark, with a grayish film.**

Fig. 92. *Sesuvium maritimum* X 19

PORTULACACEAE Purslane Family

Seeds small, lenticular or lenticular-reniform, smooth and shiny or rough-surfaced and dull; embryo curved peripherally around hard, semitransparent endosperm.

CALANDRINIA Redmaids (pl. 73)

Seeds oval, mainly thin-lenticular; 0.6–1.5 mm. across; **black, shiny, with fine cellular reticulations in a band near the margin;** often slightly oblique and with a small notch on the margin. *C. maritima* is distinct in being papillose-hairy, dull, and comparatively thick.

Fig. 93. *Calandrinia caulescens* X 10

MONTIA Minerslettuce (pl. 74)

Seeds oval-lenticular; 1–2 mm. across; black, shiny, and usually **finely cellular-reticulate on whole surface**; a **marginal cleft is plugged with white material.** *C. perfoliata* has a particularly conspicuous white plug.

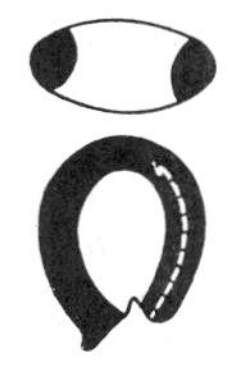

Fig. 94. *Montia perfoliata* X 10

PORTULACA Purslane (pls. 75, 76)

Seeds compressed; circular to ovate, slightly reniform; 0.5–1 mm. across; commonly with **concentric rows of low, knobby tubercles;** dark gray to black.

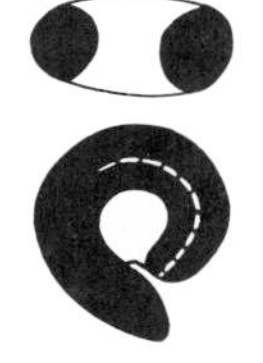

Fig. 95. *Portulaca oleracea* X 19

CARYOPHYLLACEAE Pink Family

Seeds small to medium-sized; generally **compressed and somewhat reniform** because of the peripherally curved embryo (except in *Dianthus*); surface often with **tubercles or other sculpturing in concentric rows.**

SPERGULA Spurry (pl. 77)

Globose-lenticular; 1 X 0.9 X 0.7 mm.; black, dull; with a **narrow, light-colored encircling rim or flange,** and with **whitish papillae** scattered over most of the surface. *S. maxima* is larger.

STELLARIA Chickweed (pl. 78)

Ovate-oval and **flattish-reniform or lenticular**; small, about 1 mm. across; grayish, reddish, or purplish brown; and with **concentric rows of tubercles** in the common species, *S. media*, and certain others, but the surface in some species is smooth.

CERASTIUM Chickweed (pl. 79)

Angular-ovate, flat; generally **much smaller than Stellaria,** about 0.5 mm. long; **bright reddish brown;** semitransparent; sometimes tuberculate, as in *C. viscosum*.

AGROSTEMMA Cockle (pl. 80)

Large for the family, 3 X 2.5 X 1.7 mm.; ovoid-compressed and reniform; marked by conspicuous **pointed tubercles in concentric rows; black.**

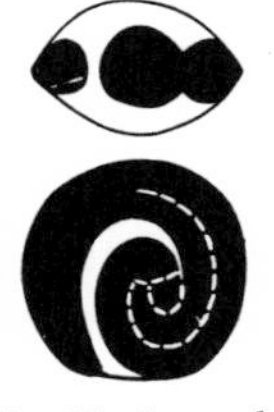

Fig. 96. *Spergula arvensis* X 14

Fig. 97. *Stellaria media* X 14

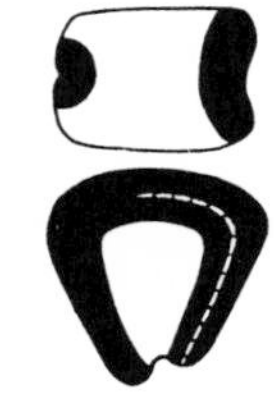

Fig. 98. *Cerastium viscosum* X 38

Fig. 99. *Agrostemma githago* X 5

SILENE Catchfly (pls. 81, 82)

Pill-like, rounded-reniform discs; 0.5–1.5 mm. across; margins ridged, rounded, or with an acute edge; surface has **tubercles in concentric rows**; dark reddish brown, black, or grayish. *Silene* and *Lychnis* are similar, but some species in each of these genera are distinct.

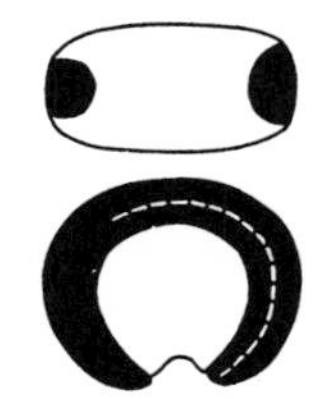

Fig. 100. *Silene latifolia* X 13

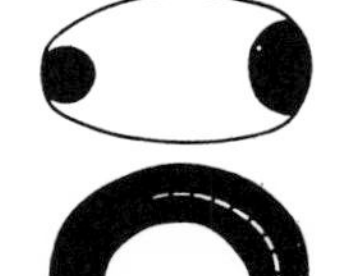

Fig. 101. *Lychnis alba* X 16

LYCHNIS Campion (pl. 83)

Very similar in shape, size, and markings to *Silene* (see above), except for *L. apetala*, which is thin-flattish, irregular, and glossy brown.

SAPONARIA Soapwort, Bouncingbet (pl. 84)

Of the two common species, *S. vaccaria* is ball-like, whereas *S. officinalis* is flat and reniform-circular. These two are similar in size (about 2 mm. across), in color (**dull black**), in surface (**with obscure pebbling**), and in internal organization.

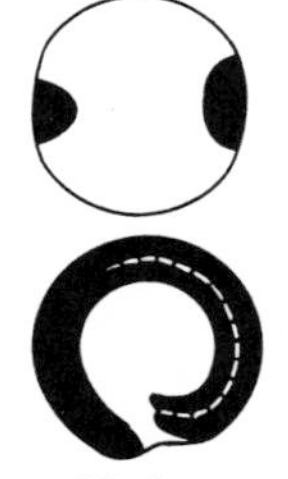

Fig. 102. *Saponaria vaccaria* X 7.5

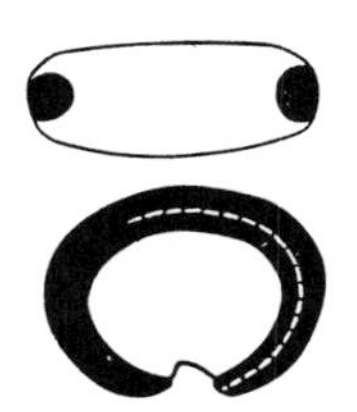

Fig. 103. *Saponaria officinalis* X 9

CERATOPHYLLACEAE Coontail Family

CERATOPHYLLUM Coontail (pl. 549)

Seed (nutlet) **elliptic-lenticular**; body about 5 X 3 X 2 mm., and with **2, 3, or more marginal spines**; wall woody, strong, with **2 layers** evident (fig. 104); endosperm lacking.

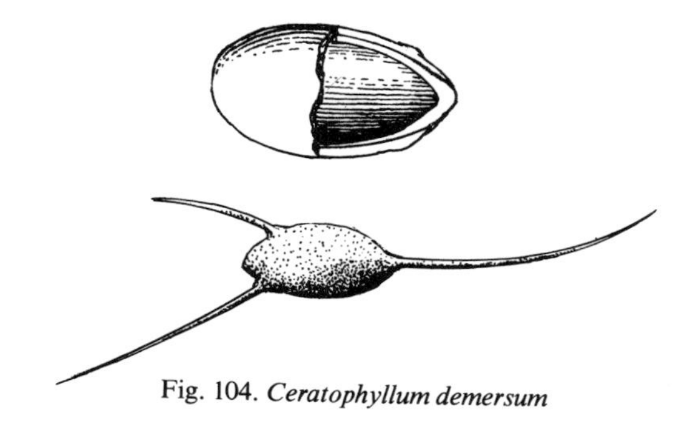

Fig. 104. *Ceratophyllum demersum*

NYMPHAEACEAE Waterlily Family

Seeds **globose to ellipsoid or obovoid**; small to large; wall delicate to very hard. The genera are similar in having a **small basal embryo and starchy endosperm**, except for *Nelumbo*, which has no endosperm.

NUPHAR Spatterdock (pl. 550)

Obovoid, **rounded on one side and compressed-angled on the other**; about 5 X 3 X 2.7 mm.; circular hole at base; yellowish brown; **wall thick** and pithy along top and one side.

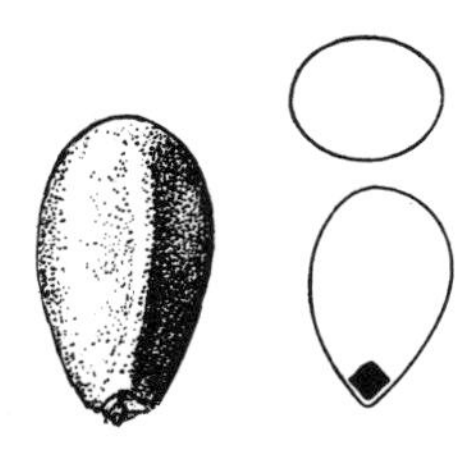

Fig. 105. *Nuphar luteum.* Diagram (right) X 3.5

NYMPHAEA Waterlily (pls. 551, 552)

Globose to ellipsoid, with a **slight ridge on one side** and a small attachment scar at the base; about 3 X 2.2 X 2 mm.; olive-green to brown; **wall fragile.**

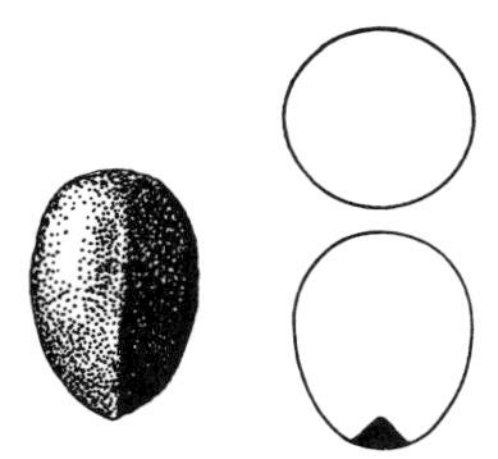

Fig. 106. *Nymphaea tuberosa.* Diagram (right) X 6

NELUMBO Lotus (pl. 553)

The large (about 1 cm. across), globose, nutlike seed **resembles an acorn,** except that it lacks the broad cup scar and has a smaller style base; the **wall is very hard**—much harder than in acorns. Endosperm is absent.

BRASENIA Watershield (pl. 554)

Globose-ellipsoid; 2.5 X 1.8 X 1.8 mm.; brownish; **wall radially striate,** woody, and about 2 mm. thick.

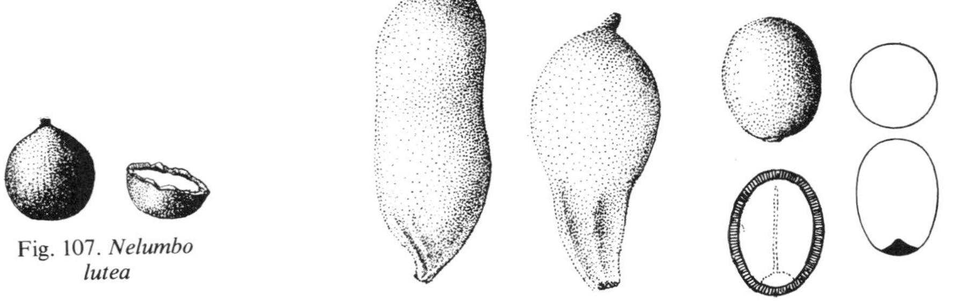

Fig. 107. *Nelumbo lutea*

Fig. 108. *Brasenia schreberi.* Diagram (right) X 6

RANUNCULACEAE Buttercup Family

Some of the seedlike structures in this family are achenes, as indicated by a persistent style base (*Ranunculus, Anemone,* etc.); the others are true seeds (*Delphinium, Aquilegia,* and many other genera). The achenes are generally flattish, whereas the true seeds are diverse in shape, size, and surface. They all have a minute to small **basal embryo** in fleshy endosperm.

RANUNCULUS Buttercup (pls. 647–652)

Diversity in size, shape, markings, and wall texture in this large genus is correlated partly with the subgenera into which it is divided. Even the embryos of different species show considerable divergence (figs. 109–111). Generally the seeds are **flattish or lenticular,** and **ovate to elliptic** in outline, **beaked** by the style base, and 0.5–2 mm. across.[12] The wall varies from papery to woody, the latter type often with finely striate markings on the inner surface. The exterior is smooth, hairy, papillate, toothed, or coarsely wrinkled.

[12] Good illustrations of nine species are included in Mason—his figures 236–244.

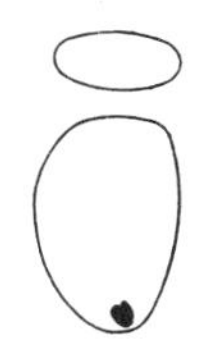

Fig. 109. *Ranunculus lapponicus* X 9

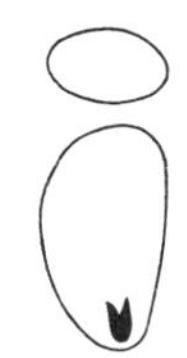

Fig. 110. *Ranunculus trichophyllus* X 15

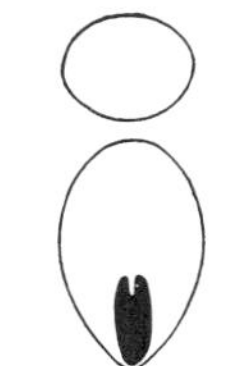

Fig. 111. *Ranunculus cymbalaria* X 30

DELPHINIUM Larkspur (pls. 85, 86)

Irregularly angled, with some faces flattened by compression; commonly 2–3 mm. long. Seeds in some species smooth and glossy, with either a tight or a loose seed coat, but in others covered with a series of ridges, scales, or wings, as in *D. virescens*.

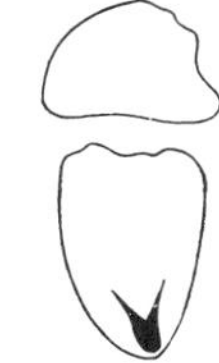

Fig. 112. *Delphinium geyeri* X 6

ANEMONE Anemone (pl. 87)

A distinctive feature of most species is the **woolly coating** of whitish hairs over the **flat,** oval to ovate seeds. In *A. caroliniana,* however, pubescence is scant, and the seeds are pink and green. Seeds of the genus vary in length from 2 to 6 mm.

Fig. 113. *Anemone caroliniana* X 10

BERBERIDACEAE Barberry Family

BERBERIS Barberry (pls. 88, 89)

Mainly **ellipsoid to oblong,** rounded at top and tapering toward base, **terete, to flattened or concave** on one or more faces; dark brown, glossy or dull; about 3–5 mm. long and 1–2 mm. thick.

Fig. 114. *Berberis aquifolium* X 3.5

MENISPERMACEAE Moonseed Family

Seeds woody, circular, and flat in *Cocculus* and *Menispermum*.

COCCULUS Snailseed (pl. 812)

Somewhat **doughnut-shaped;** 4.5 X 3.6 X 1.8 mm.; light brown or straw-colored; wall woody, **sculptured radially by low ridges and grooves.**

MENISPERMUM Moonseed (pl. 813)

Seed **circular,** similar to that of **Cocculus;** resembles a **scallop-margined pie** with a large wedge removed; 7 X 6 X 2 mm.; woody.

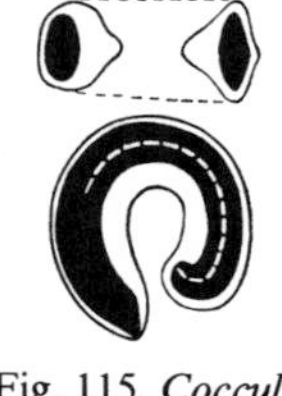

Fig. 115. *Cocculus carolinus* X 3.5

Fig. 116. *Menispermum canadense* X 3.5

MAGNOLIACEAE Magnolia Family

Seeds large or medium-sized, of various shapes, woody or bony, uniform in having a small basal embryo.

MAGNOLIA Magnolia (pls. 705, 706)

Large, about 1 cm. long; **oval to elliptic, biconvex,** often with either a low lengthwise ridge or **shallow groove** in the middle of a face and a **small hole** at the broader end; wall somewhat bony.

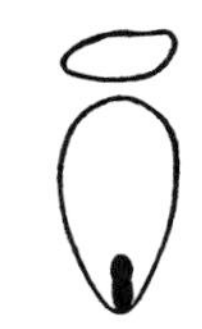

Fig. 117. *Magnolia acuminata* X 2

LIRIODENDRON Tuliptree (pl. 704)

Seeds borne in the woody base of an ashlike samara; compressed, elliptic to oval, **pointed at top** and with a small, **roundish hole at base;** 5 X 2.5 X 1.5 mm.; surface **warty-rough.**

Fig. 118. *Liriodendron tulipifera* X 4

LAURACEAE Laurel Family

SASSAFRAS Sassafras (pl. 707)
LINDERA Spicebush (pl. 708)

The seeds of sassafras and spicebush are similar in being **ellipsoid to globose,** with a dark, **brittle, papery wall,** and having a short embryo stalk buried in a groove between 2 thick cotyledons (no endosperm). They can be distinguished readily by: (*a*) shape—sassafras is shorter, 6 X 5 X 4.5 mm., as compared to spicebush, which is 7.5 X 4.5 X 4.5 mm.; (*b*) markings—sassafras is merely dark or

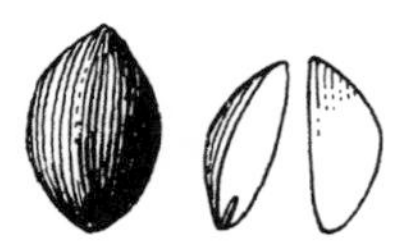

Fig. 119. *Sassafras officinale*

black, whereas spicebush is **mottled with grayish;** (*c*) wall thickness—the wall of sassafras is at least twice as thick as that of spicebush; (*d*) embryo—the embryo stalk of spicebush is at least 2 mm. long, whereas that of sassafras is generally less than 2 mm. long. The closely related genus *Persea* is more uniformly globose.

PAPAVERACEAE Poppy Family

Seeds small; globose to roundish-reniform; surface mainly reticulate, but smooth in some; embryo basal, minute to medium small.

ARGEMONE Pricklypoppy (pl. 90)

Nearly globose; about 2 mm. across; with **2 puckered-up, hornlike points on opposite sides,** connected by a narrow ridge; surface dark or grayish, coarsely reticulate, mainly with pentagonal cells.

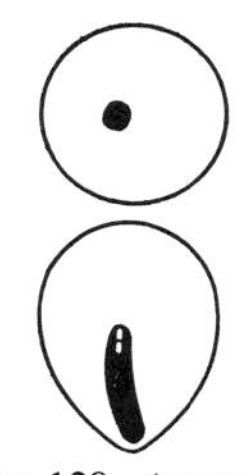

Fig. 120. *Argemone platyceras* X 10

ESCHSCHOLTZIA California-poppy (pl. 91)

Globose or nearly so; slightly smaller than *Argemone,* and lacking the horns; **a dark, narrow line extends between seed poles;** surface **coarsely reticulate,** with delicate, readily removed ridges; dark or grayish; about 1.8 mm. across.

Fig. 121. *Eschscholtzia californica* X 9

PAPAVER Poppy (pl. 92)

Small, about 1 mm. long; roundish-reniform or like a short, thick, arched rod; with **coarse reticulations** aligned in rows around the arched back; dark reddish purple or creamy brown; embryos not uniform in size in different species (figs. 122, 123).

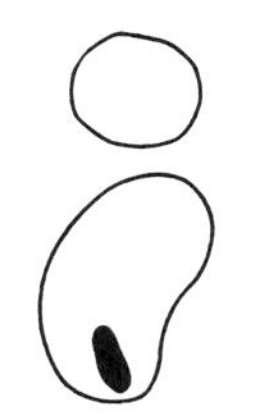

Fig. 122. *Papaver dubium* X 22

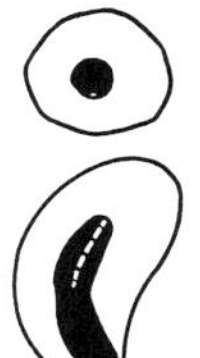

Fig. 123. *Papaver somniferum* X 13

CAPPARIDACEAE Caper Family

Seeds compressed-globose to ellipsoid, with contours on the exterior showing that the embryo is curled or arched over on itself; diverse in size; surface rough to smooth; color light to dark; endosperm scant, fleshy.

POLANISIA Clammyweed (pl. 93)

Compressed-globose; 2.2 X 1.9 X 1.1 mm.; **dark purplish;** with or without easily removed, delicate tubercles under which the surface is **finely reticulate in curved lines;** narrow line from marginal notch extends beyond middle of seed.

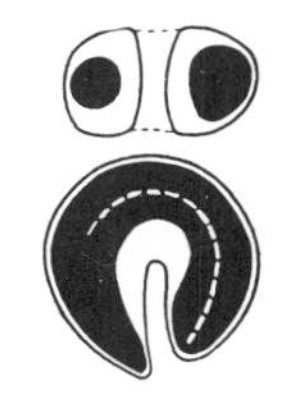

Fig. 124. *Polanisia graveolens* X 7

CLEOME Spiderflower (pls. 94, 95)

Similar to *Polanisia* (above), but **more compressed** and faces flatter; **tubercled;** size more variable, 2–3 mm. across; color in mature specimens blackish or brown-black.

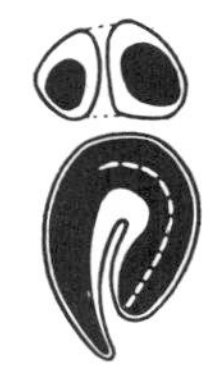

Fig. 125. *Cleome serrulata* X 5

CRUCIFERAE Mustard Family

Seeds of various shapes, from globose or ellipsoid to rodlike, flat or winged, generally with **contours showing that the embryo is bent or folded;** surface commonly finely reticulate-pitted.[13] This is the only large family having folded embryos and lacking endosperm in significant amounts.

LEPIDIUM Pepperweed (pls. 96, 97)

Mainly flat, oval to elliptic, and rusty brown; but a few, such as *L. campestre* and *L. draba.* are ovoid to ellipsoid and black or reddish black. In most of the species characterized by flat seeds, the seed has one straight, thick edge and is **thin or narrowly winged on the other edge.** About 2 mm. long. *L. perfoliatum* is winged all the way around.

CAPSELLA Shepherds-purse (pl. 98)

Compressed-elliptic; **minute,** 1 X 1.2 X 0.33 mm.; light brown, except for a **darker zone across the attachment point.**

---

[13]Good drawings of seeds of this family, as well as keys, are included in Seeds of the Cruciferae of Northeastern North America, by Margaret R. Murley, Am. Midl. Nat. 46:1–81, 1951.

BRASSICA Mustard (pls. 100, 101)

Spheroid to irregularly globose; 1.5–2 mm. across; dark reddish brown or purplish brown to black; **generally finely reticulate;** a small attachment scar is evident, and in some species, especially in immature material, there is a ridge bordered by 2 shallow grooves on one side. Identification characters of this genus are discussed and illustrated in a Department of Agriculture publication by Albina Musil.[14]

SISYMBRIUM Tumblemustard (pl. 99)

Oblong-truncate, somewhat wedge-shaped; one side with two flattish faces joined at a low angle and the other side flat; a **shallow notch** at the attachment end; about 1 mm. long.

BARBAREA Wintercress (pls. 102, 103)

Oblong-oval, compressed, somewhat truncate and **wedge-shaped;** with a small attachment notch; 2 X 1.2 X 0.8 mm. in *B. verna* and 1.3 X 1 X 0.6 mm. in *B. vulgaris;* surface scurfy-reticulate, grayish.

ARABIS Rockcress (pls. 105, 106)

Flat, circular, oval, or angled-oval; generally 1–2 mm. across; with an **encircling marginal wing** of variable extent; embryo stalk evident along the inner edge of the broad cotyledons.

DESCURAINIA Tansymustard (pl. 104)

Minute, 0.8–1.5 X 0.5 X 0.5 mm.; oblong-oval, compressed or terete; brown; generally **finely cellular-reticulate in lengthwise lines;** with a white attachment point bordered by a contrasting dark zone.

HAMAMELIDACEAE Witch-hazel Family

No external characters serve consistently to distinguish the seeds of this family; internally, however, they are similar in having **spatulate embryos** in fleshy endosperm.

HAMAMELIS Witch-hazel (pl. 710)

Ellipsoid; 8 X 5 X 4 mm.; **glossy black,** with a large, light-colored attachment scar on the truncate basal area; wall woody, about 0.5 mm. thick, with obscure fibers running crosswise on the inner face.

Fig. 126. *Hamamelis virginiana* X 2.5

[14]Distinguishing Species of *Brassica* by Their Seeds, by Albina F. Musil, U. S. Dept. Agr. Misc. Publ. 643, 1948.

LIQUIDAMBAR Sweetgum (pl. 709)

Terminally winged; 8 X 3 X 1 mm.; with **light and dark streaks** along fine lengthwise lines radiating from base of concave side.

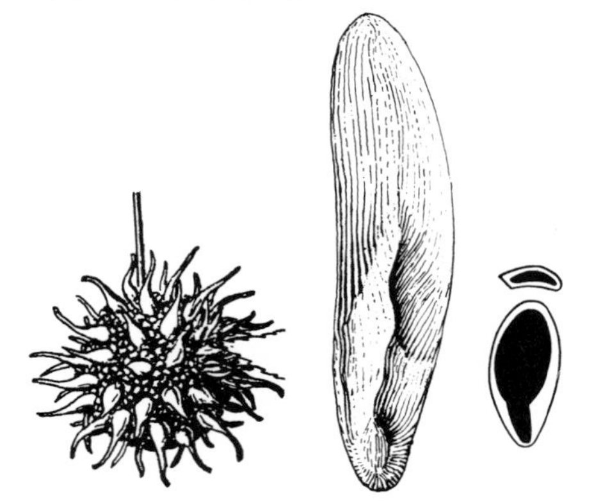

Fig. 127. *Liquidambar styraciflua.* Diagram (right) X 3

PLATANACEAE Sycamore Family

PLATANUS Sycamore (pl. 711)

Unique in being **club-shaped,** with a **tuft of stiff, buffy hairs** pointed upward from the base; body 8–9 X 1–1.5 X 1 mm.; light brown, with a persistent style base. Without their hairs, sycamore seeds are remarkably similar to those of buttonbush (pl. 795), but the latter are broader, shorter, and 2-celled inside.

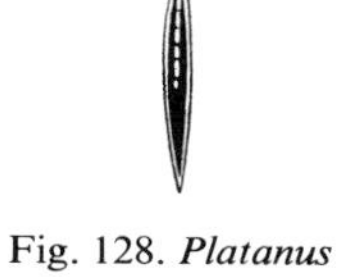

Fig. 128. *Platanus occidentalis* X 3

ROSACEAE Rose Family

Seeds of this large, diverse family have no consistent identifying features, partly because some of them are achenes, others are pits of drupes, and many others are true seeds borne in follicles or fleshy fruits. Endosperm little or none; most of the interior occupied by large cotyledons, which, in some species, embrace the short embryo stalk.

MALUS Apple (pl. 712)
PYRUS Pear (pl. 713)

Similarities in the compressed-ovate to compressed-elliptic seeds of apples and pears support their classification under a single genus, *Pyrus.* However, seeds of cultivated apples can be distinguished from those of pears fairly readily by the fact that **pear seeds incline to be more attenuate, and their surface lacks the fibrous lengthwise striations of apple seeds.** Endosperm is negligible in both, except that it is present in small amounts in some species of *Malus* (fig. 129).

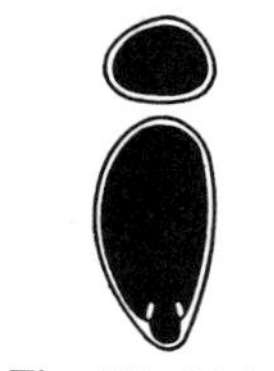

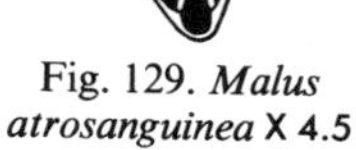

Fig. 129. *Malus atrosanguinea* X 4.5

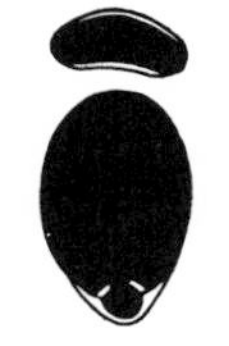

Fig. 130. *Photinia arbutifolia* X 5

PHOTINIA Toyon (pl. 714)

Plano-convex; often with a **low ridge running lengthwise** on the plane surface; purplish brown or reddish brown, dull, with obscure cellular reticulations; 3–4 X 2–3 X 1.5 mm.

AMELANCHIER Serviceberry (pl. 717)
SORBUS Mountain-ash (pl. 716)
ARONIA Chokeberry (pl. 715)

Seeds of these three genera are similar in size (3–5 mm. long) and in their shape, which is elliptic-ovoid with a somewhat curved or **hooked base.** One edge is rounded, and the other often nearly straight or incurved and comparatively thin. Generally, however, *Amelanchier* is **dark and large,** Sorbus is usually **bright brown and comparatively slender,** and *Aronia* is **small and dark,** with the surface usually roughened by low lengthwise ridges.

Fig. 131. *Amelanchier canadensis* X 4

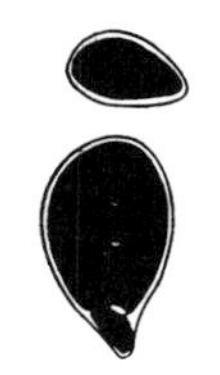

Fig. 132. *Sorbus aucuparia* X 4.5

Fig. 133. *Aronia arbutifolia* X 6.5

COTONEASTER Firethorn (pl. 718)

Seeds (nutlets) woody, **rounded-triangular** with two flat faces and a rounded back on the common species, *C. pyracantha*, or **plano-convex** in larger-seeded species; 2.5–5 mm. long. The **plane faces are partly or entirely glossy,** as if varnished, whereas the rounded side is partly or entirely rough and dull-surfaced.

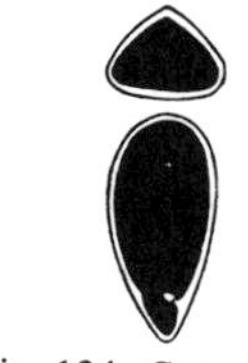

Fig. 134. *Cotoneaster pyracantha* X 7.5

CRATAEGUS Hawthorn (pls. 719, 720)

Seeds (nutlets) mainly **rounded-triangular,** with two plane faces and a rounded back on which there usually are **coarse ridges and grooves;** 5–10 mm. long; commonly there is a **small hole or notch** at one end of the angular edge; wall thick and hard; frequently the nutlets are sterile.

Fig. 135. *Crataegus mollis* X 3

ROSA Rose (pls. 724, 725)

Seeds (nutlets) **rounded-triangular** and often oblong, usually with a rounded back and two flattish faces joined into an edge bearing a **shallow, narrow groove;** yellowish to brown; 3–5 mm. long; wall woody, thick, its **inner surface lined by a minutely pitted, glossy layer,** in which **fibers run crosswise** and frequently are loose along a cut or break in the wall.

Fig. 136. *Rosa bracteata* X 5

RUBUS Blackberry, Raspberry (pls. 721–723)

Seeds (nutlets) various, but generally compressed, ovate or oval-oblong, with a **straight edge** and a broader, rounded back; faces and back **coarsely reticulate,** with a **narrow groove on the straight edge** and a **low ridge on the back;** usually 2–3 mm. long.

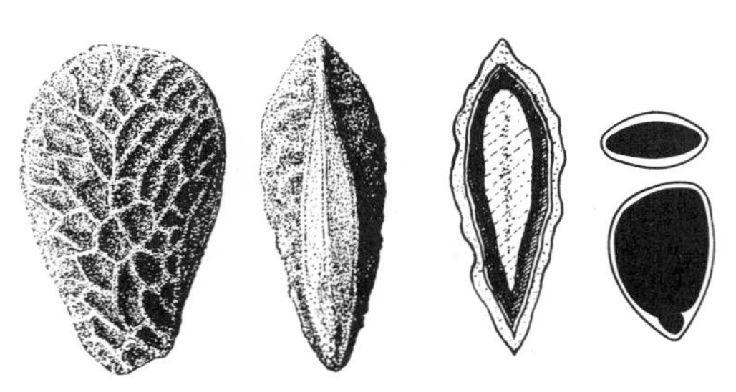

Fig. 137. *Rubus cuneifolius.* Diagram (right) X 10

PRUNUS Cherry, Plum (pls. 726–730)

Plum stones are generally **compressed**, oval to elliptic, whereas cherry pits are more globose and vary from **compressed-oval** to **spheroid.** Both have a low ridge along one side; a section through this shows a **channel in the wall, near the outer edge.** The inner surface is **minutely cellular-pitted.** Cherry pits often show a hole at their attachment end.

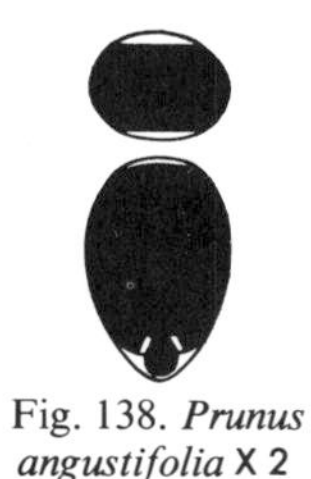

Fig. 138. *Prunus angustifolia* X 2

FRAGARIA Strawberry (pl. 107)
POTENTILLA Cinquefoil (pls. 108, 109)

Seeds of these two genera are similar in size and shape, usually 1–2 mm. long, ovate- or elliptic-compressed, frequently arched, with a rounded back and a straight or concave edge. Most species have **fingerprint-like sculpturings on the surface.** Strawberry seeds are distinct in being broader, particularly at the large end, owing to the outward projection of their attachment area.

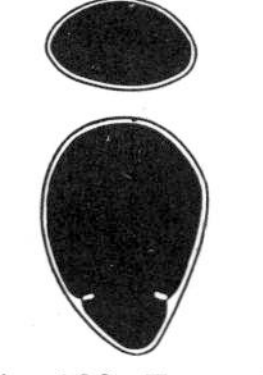

Fig. 139. *Fragaria chiloensis* X 14

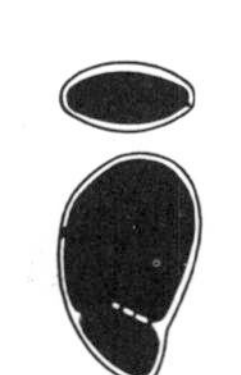

Fig. 140. *Potentilla monspeliensis* X 19

GEUM Avens (pl. 110)

Seed (achene) body elliptic to oblanceolate; usually 2–3 mm. long; dark brown; hairy; with a **long, rigid style tipped by a hook.** Endosperm lacking.

AGRIMONIA Agrimony (pl. 111)

Globose seed (achene) about 2 mm. across, concealed in a **top-shaped** covering (hypanthium), the "equator" of which is **armed by numerous hook-tipped bristles.** Endosperm none.

LEGUMINOSAE Bean Family

Many legume seeds are characteristically beanlike, having a compressed-oval shape and a conspicuous, notched attachment area (hilum) near the middle of one edge. Many others, however, are different. Some are rodlike, rectangular, or globose, and in some the attachment scar is obscure. Usually the **seed coat is thin, smooth, and tough,** and **in some species it is mottled, streaked, or spotted.** The interiors consist mainly of broad, thick cotyledons, but in some the endosperm is extensive and the cotyledons are relatively small.

ACACIA Acacia (pls. 731, 732)

Rather diverse: compressed and oval to elliptic; black to greenish brown; 4–12 mm. long. All species have the faces marked by an oval or elliptic line more or less concentric with the outline of the seed. Endosperm lacking.

PROSOPIS Mesquite, Screwbean (pls. 733, 734)

Compressed and oval, ovate, or elliptic; 2.5–7 mm. long; brown, rather glossy; with a central ring on each face—a feature shared by several other genera. Cross sections show the **yellowish embryo sandwiched between 2 hard, glassy layers** of endosperm.

Fig. 141. *Prosopis pubescens* X 5

SCHRANKIA Sensitivebrier (pl. 116)

Seeds similar to those of some *Desmanthus* species; rhomboid-oblong and dark, or compressed-elliptic and brown; faces marked by a raised, oblong, central zone.

Fig. 142. *Schrankia uncinata* X 3.5

DESMANTHUS Bundleflower (pls. 112, 113)

Diverse in different species; *D. illinoensis* is flattish-ovate, 3.5 X 2.5 X 1 mm., with an oval to elliptic central zone; *D. leptolobus* is similar but narrowly elliptic, 5 X 2 X 1 mm.; other species are somewhat rhomboid, with a hump on each face and bearing a **crescent-like mark.** Embryo yellowish, sandwiched between layers of hard, semitransparent endosperm.

Fig. 143. *Desmanthus illinoensis* X 4.5

MIMOSA Mimosa (pl. 735)

Compressed and oval to elliptic; brown to gray; about 4 X 2.5 X 1.5 mm., except for *M. pigra,* which is 6 mm. long. The latter and *M. pudica* have a **central ring** on each face, whereas other species are marked by a **dark crescent on the humplike central area.**

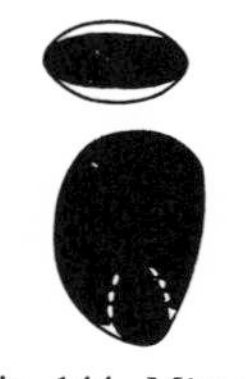

Fig. 144. *Mimosa biuncifera* X 5

CASSIA Senna (pls. 114, 115)

Diverse externally and internally (figs. 145–147); medium small to large; compressed-oval, flattish-circular, flattish-spatulate, or rhomboid. All species have faces marked by an **elliptic to circular zone** or ring.

Fig. 145. *Cassia tora* X 3.5

Fig. 146. *Cassia ligustrina* X 4.5

Fig. 147. *Cassia occidentalis* X 4

CHAMAECRISTA Partridgepea (pls. 117, 118)

Distinctive in their **flattish-rhombic or lozenge shape and rows of circular punctations.** The two common Eastern species *C. nictitans* and *fasciculata* are **black,** the former more glossy and smaller than the latter. *C. leptadenia* of the Southwest is brownish.

Fig. 148. *Chamaecrista nictitans*

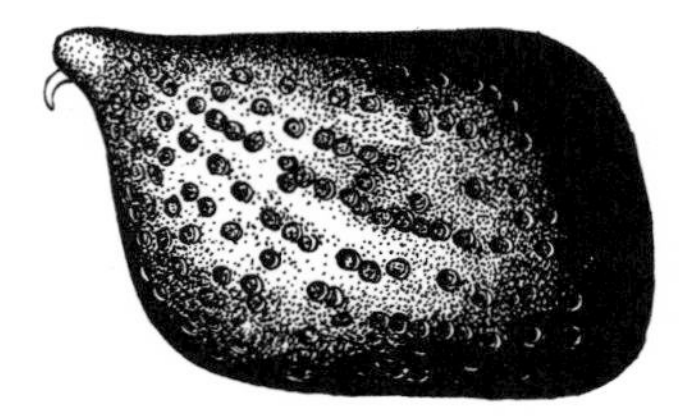

Fig. 149. *Chamaecrista fasciculata.* Diagram (right) X 5

CROTALARIA Crotalaria (pls. 119, 120)

Fairly uniform in shape but diverse in size; 2–8 mm. long; compressed or flattish and **asymmetric-reniform**, with one end arched over only part way toward the other end; often with a **curved beak and a prominent notch;** glossy to dull, yellowish, brown, or black.

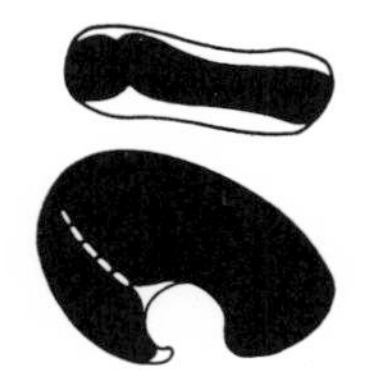

Fig. 150. *Crotalaria spectabilis* X 5

LUPINUS Lupine (pls. 121–124)

Distinct in having its **attachment scar** (hilum) **terminal,** and often surrounded by a collar which protrudes outward; generally compressed-ovoid; 2.5–10 mm. long; solid creamy or black in a few species, but more commonly mottled or streaked with dark and light markings; often with a **squarish lighter-colored zone** at one side of the hilum area; endosperm none.

MEDICAGO Burclover, Alfalfa (pls. 125, 522)

Except for *M. lupulina,* which is compressed-ovoid, *Medicago* seeds are **compressed-falcate,** like a new moon with the ends rounded; from the hilum area, a straight, **narrow line runs obliquely** on each face; yellow to brown, dull; 2–3 mm. long. Alfalfa (*M. sativa*) seeds are often truncate near the ends; *M. arabica* has a distinctive point projecting outward from its hilum area.

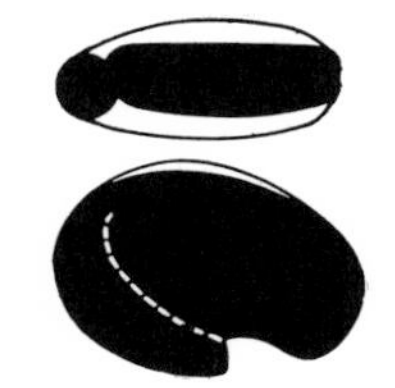

Fig. 151. *Medicago lupulina* X 9

TRIFOLIUM Clover (pls. 126–130)

Either compressed-ovoid to ovoid, with a **small, circular hilum** flush with the surface, or compressed-ovoid and truncate at one end, with the hilum in a depression; seeds various in size, 1–3 mm. long; yellow, brown, black, or mottled; glossy to dull.

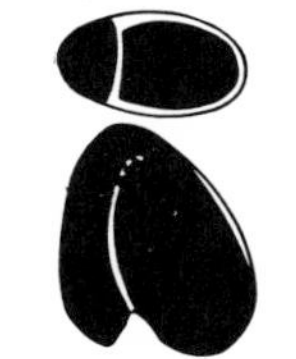

Fig. 152. *Trifolium repens* X 9

LOTUS Deervetch (pls. 131–135)

Extremely **diverse in shape,** from ovoid or spheroid (*L. tomentellus*), rodlike (*L. micranthus, scoparius*), or compressed-oval, to rhomboid; generally small but ranging from about 1 mm. to 4 mm. in length; **hilum small, circular;** usually mottled brown, greenish, or black, on a gray or creamy background; glossy to dull or rough-surfaced.

Fig. 153. *Lotus crassifolius* X 6

MELILOTUS Sweetclover (pls. 136, 137)

Compressed-oval, with a **small, circular hilum** on an edge near one end. *M. alba* is somewhat truncate at the broader end, is yellow, and has a distinct line running lengthwise from the hilum; other species are brown or greenish brown; *M. officinalis* shows the lengthwise line only obscurely.

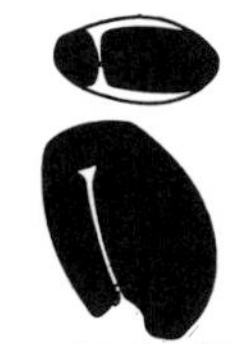

Fig. 154. *Melilotus alba* X 9

PSORALEA Scurfpea (pl. 138)

Compressed, oval to ovoid; 3–6 mm. long; dark-colored or streaked and mottled (*P. macrostachya*); **hilum oval, white** or light-colored, generally with a low **ridge or hump** near one end; hilum not quite median on seed edge; endosperm none.

Fig. 155. *Amorpha canescens* X 6

AMORPHA Leadplant (pl. 736)

Compressed, and oval or elliptic to lanceolate; 2–5 mm. long; **circular hilum near the end of the nearly straight edge,** along which a narrow dark line runs lengthwise; end of seed somewhat arched near hilum; brown, glossy or dull.

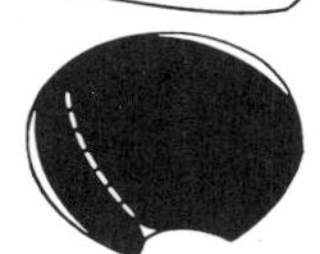

Fig. 156. *Indigofera caroliniana* X 7.5

INDIGOFERA Indigo (pl. 139)

Varying from **oblong block-shaped or truncate-cylindric** (like sawed-off segments of a solid cylinder) to ovoid or spheroid; about 2 mm. long; **hilum circular, nearly central on one margin;** olive-green, brown, black, or mottled; *I. caroliniana* is distinct in its black-eyed hilum.

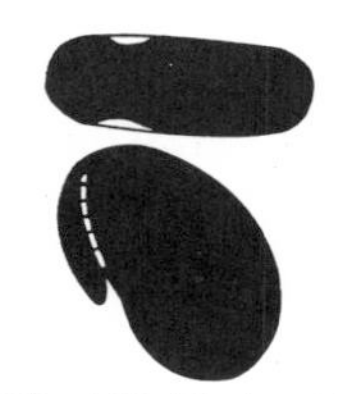

Fig. 157. *Tephrosia virginiana* X 6

TEPHROSIA Hoarypea (pl. 140)

Mainly compressed and oval to oblong, often truncate at one end; 2–6 mm. long; hilum marginal, not quite central, with a **white, papery collar;** in several species, black mottling contrasts with grayish or brownish background.

ROBINIA Black Locust (pl. 737)

Compressed, and elliptic or oval, somewhat **falcate;** fairly large hilum in a notch beside a **prominent hump; white collar** around the hilum; mottled black on gray or on brown; about 5 X 3 X 1.5 mm.

SESBANIA Coffeeweed (pl. 143)

Oblong-ellipsoid and slightly compressed; 4 X 2.5 X 2 mm.; blackish mottling on gray or brown; hilum circular, nearly central.

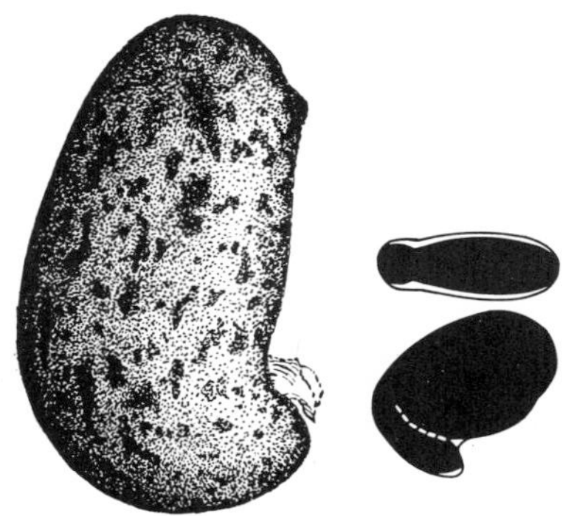

Fig. 158. *Robinia pseudoacacia.* Diagram (right) X 4

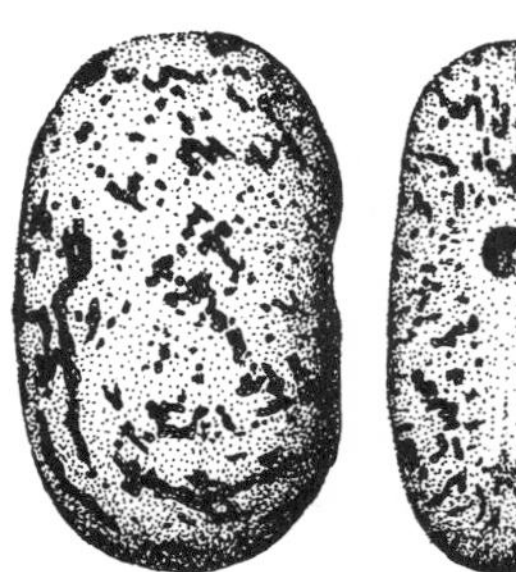

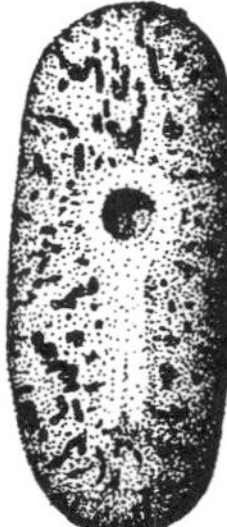

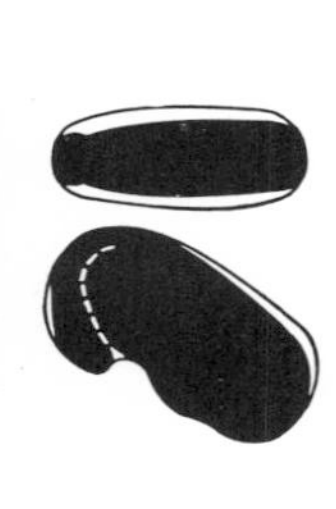

Fig. 159. *Sesbania macrocarpa.* Diagram (right) X 4.5

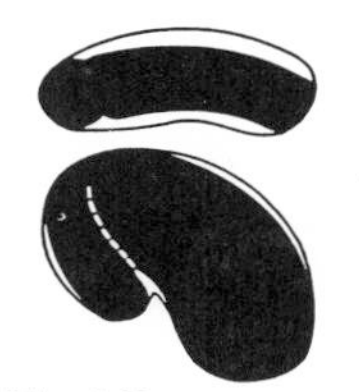

ASTRAGALUS Loco (pls. 141, 142)

Diverse: **flattish,** and oval, ovate, circular, or truncate; usually with the **minute hilum in a deep marginal notch;** mainly 2–3 mm. long; brown to black or mottled.

Fig. 160. *Astragalus brazoensis* X 7.5

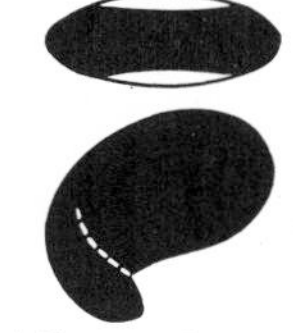

AESCHYNOMENE Jointvetch (pl. 144)

Seeds (borne in rectangular loment segments) flattish-compressed, **oval, with a broad, deep hilum notch** and a large, oval hilum, which is cleft lengthwise and generally is adjoined by a hump; 4 X 3 X 1.5 mm.; dark brown.

Fig. 161. *Aeschynomene virginica* X 5

DESMODIUM Tickclover (pls. 145–147)

Seeds (borne in oval or oval-triangular loment segments coated with hooked hairs) flattish to compressed and oval or ovate to elliptic or somewhat triangular (*D. nudiflorum*); 2–8 mm. long; brown; **hilum small, circular, nearly central, on an almost straight edge;** one end of seed usually broader than the other.

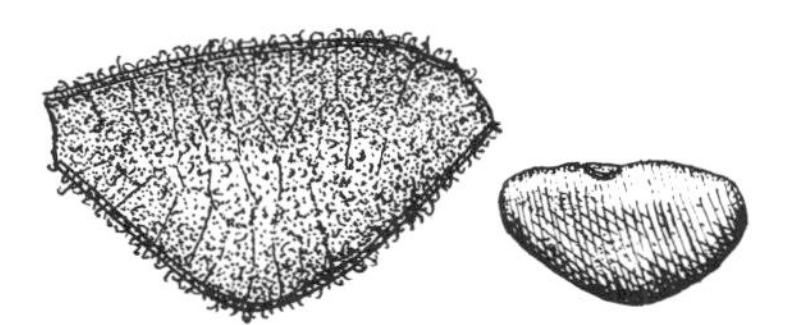

Fig. 162. *Desmodium dillenii*

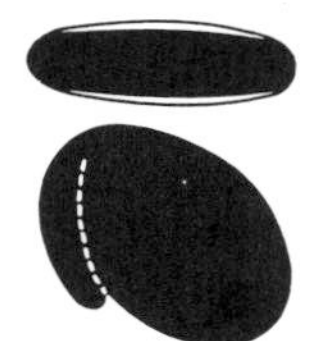

Fig. 163. *Desmodium tortuosum* X 7.5

LESPEDEZA Lespedeza (pls. 148–152)

Seeds (borne in a pointed pod partially enclosed by the calyx) similar in shape to *Desmodium,* but proportionately **thicker, not as straight on the hilum edge,** and the **circular, white-lined hilum definitely not central;** usually small, 1.5–4 mm. long; green, brown, or black, or mottled. Seeds of native species are generally green or nearly black. The two important annual species can be distinguished readily as follows:

Lespedeza striata (Common Lespedeza)
- Pod acuminate at top.
- Calyx large, brownish, covering much of pod.
- Seed mottled, purplish brown.
- Hilum in a slight notch, on edge near end.

Lespedeza stipulacea (Korean Lespedeza)
- Pod broad-obtuse at top.
- Calyx small, semitransparent, covering only basal one-fourth of pod.
- Seed solid black or purple.
- Hilum not in notch, and practically on end.

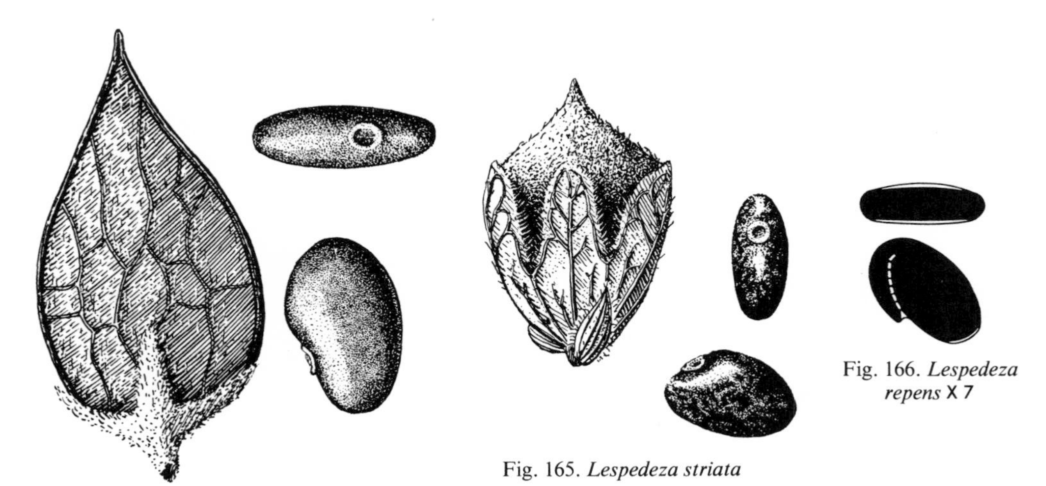

Fig. 164. *Lespedeza frutescens*

Fig. 165. *Lespedeza striata*

Fig. 166. *Lespedeza repens* X 7

VICIA Vetch (pls. 153, 154)
LATHYRUS Peavine (pls. 155, 156)

These two genera cannot be distinguished satisfactorily by their seeds; perhaps they should be united. Their seeds are mainly globose or compressed-globose, 1.5–7 mm. across (*V. faba* is 15 mm. long), cream, brown, or black, with an **oval, oblong, or linear hilum.** Endosperm none.

APIOS Groundnut (pl. 814)

Rounded rectangular-solid, with irregular or wrinkled surface; 6 X 5 X 4.5 mm.; dark brown; hilum inconspicuous, often with a small, brown, papery attachment at its edge. Endosperm none.

GLYCINE (SOJA) Soybean (pl. 523)
VIGNA Cowpea (pl. 524)

Seeds of both of these genera vary from compressed-oval to spheroid; white or brown to black; and average about 8 mm. in length. They can be distinguished from each other readily by the hilum, which in *Glycine* is **oblong, flush** with the surface, and marked by a **central lengthwise line** or cleft (much as in *Vicia* and *Lathyrus*), whereas in *Vigna* the hilum is generally **oblong-wedge-shaped** and is covered with **horny-appearing material** which protrudes slightly above the seed surface. There is a cavity between the cotyledons in *Vigna* but not in *Glycine.* Endosperm none.

CENTROSEMA Butterflypea (pl. 157)

Mainly oblong-solid, with rounded margins and only slightly compressed; 3–4 X 2.5 X 2 mm.; with a large (nearly one-third of the seed length) oval-elliptic hilum surrounded by a **dark marginal band,** the outer edge of which is rather irregular, in contrast with the regular outer edge of the band on *Galactia*; reddish brown, with black, irregular or oblong streaks and mottlings; no endosperm.

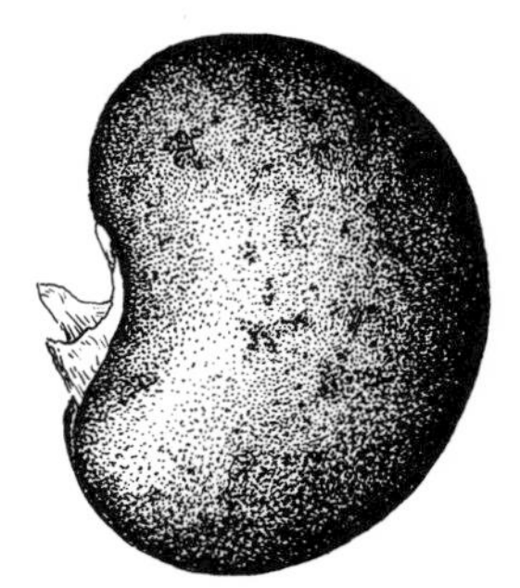

Fig. 167. *Amphicarpa bracteata*

AMPHICARPA Hogpeanut (pl. 815)

Broadly compressed-oval, with one **straightish side;** 3.5 X 3 X 1.5 mm.; hilum oblong, about 1.5 mm. long, whitish, and with a persistent **papery collar;** dark reddish brown with black streaks and mottlings; no endosperm.

GALACTIA Milkpea (pl. 162)

Slightly compressed, **oblique-ovoid or oblique-truncate;** in the common species 2–3.5 mm. long, but in some species, longer; hilum conspicuous, with a **broad, dark collar** which (in contrast to the collar in *Centrosema*) is **clearly and regularly outlined** on its outer margin as well as on its inner one; usually marked by lengthwise **dark, mottled bands alternating with lighter ones,** or streaked in other ways. No endosperm.

STROPHOSTYLES Wildbean (pls. 159–161)

Oblong-solid, more or less compressed; 3–8 mm. long; outer coating grayish scurfy, and black underneath (*S. helvola, umbellata*), or surface smooth and mottled, either brownish or grayish; hilum oblong, covered by white material and bordered by a narrow, black collar. Cavity between cotyledons, as in *Vigna*; no endosperm.

Fig. 168. *Galactia volubilis*

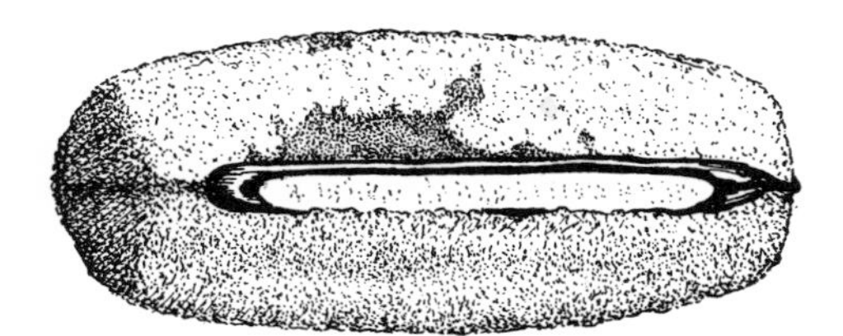

Fig. 169. *Strophostyles helvola*

RHYNCHOSIA Rhynchosia (pl. 158)

Compressed or **flattish, circular to elliptic,** with **one edge straight or concave;** hilum oval, covered by papery material; reddish brown or grayish and mottled. No endosperm.

PHASEOLUS Bean (pls. 163, 525)

Diverse in shape, size, and color; compressed-oval, -elliptic, or -lunate, to globose; 4–20 mm. long; white, black, brown, reddish brown, green, or mottled; **hilum elliptic to linear, covered with whitish material.** No endosperm.

LINACEAE Flax Family

LINUM Flax (pls. 164, 526)

Seeds borne in flattish, thin-walled compartments like narrow sectors of a sphere; flat or flattish, elliptic or elliptic-ovate, one side generally nearly straight. Native species 1–2 mm. long; light-colored. Cultivated flax, *L. usitatissimum,* 4 mm. long; brown, glossy.

Fig. 170. *Linum berlandieri* X 5

OXALIDACEAE Oxalis Family

OXALIS Oxalis (pl. 165)

Flat, oval-ovate; sculptured with cross ridges, except in *O. acetosella,* which has longitudinal ridges; **reddish brown;** generally slightly more than 1 mm. long, but *O. acetosella* is about 3 mm. long.

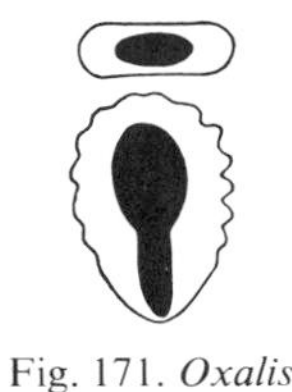

Fig. 171. *Oxalis stricta* X 14

GERANIACEAE Geranium Family

Seeds mainly small; ellipsoid to rodlike; **embryo folded;** endosperm negligible.

GERANIUM Geranium (pls. 166, 167)

Ellipsoid or angular-ellipsoid; often sculptured with a meshwork of cellular reticulations; narrow line from middle of one end extends just over edge; one side often somewhat ridged, with adjoining faces flattish or concave; when outer reticulated coating is removed, an inner glossy brown coat is seen; embryo greenish or yellow, folded; endosperm none or negligible.[15]

Fig. 172. *Geranium carolinianum*

ERODIUM Filaree (pls. 168, 169)

Oblanceolate-cylindric; 3–5 mm. long; a fine line extends from narrowed end up the side nearly 1 mm.; removal of the dull, brownish outer wall reveals an inner glossy brown one; embryo folded; endosperm negligible.[16]

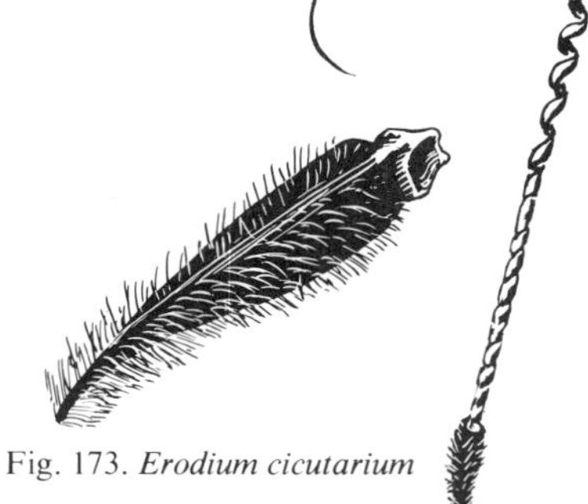

Fig. 173. *Erodium cicutarium*

[15]See A Seed Key to Fourteen Species of Geraniaceae, by Margaret R. Murley, Proc. Iowa Acad. Sci. 51:241–246, 1944.

[16]See Alfileria (Filaree) Seed, by William A. Dayton, Rhodora, 39:323–325, 1937.

ZYGOPHYLLACEAE Caltrop Family

No consistent family characters in seeds, either externally or internally.

TRIBULUS Puncturevine (pl. 170)

Seed (hard, dry fruit) distinctive in **its 2 tough spines,** about 5 mm. long, arising obliquely from a reticulate-ridged, **irregular base.** Endosperm none.

Fig. 174. *Larrea tridentata* X 4.5

LARREA Creosote (pl. 738)

Crescent-shaped, compressed, with the incurved margin narrow; surface rough; brown to black; about 3.5 mm. long.

KALLSTROEMIA Caltrop (pl. 171)

Seeds (carpels) woody; **flat and broadly wedge-shaped;** the flat, glossy, **reticulate-sculptured** faces sloping into a thin, straight edge; the other (rounded) edge is thicker and bears a series of blunt, **hornlike tubercles,** the bases of which are covered with short, whitish hairs; 4 X 3 X 2 mm.; endosperm none.

POLYGALACEAE Milkwort Family

POLYGALA Milkwort (pls. 172, 173)

Generally **pubescent**, and with a papery, **whitish aril**. Two main types of seeds: one small (0.5–1.5 mm. long), ovoid, and only moderately hairy; the other larger (4–5 mm. long), truncate-oblanceolate, flattish, and densely coated with pubescence.

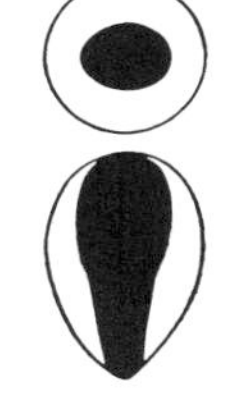

Fig. 175. *Polygala sanguinea* X 15

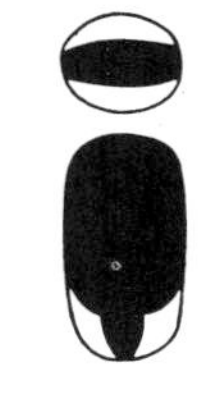

Fig. 176. *Polygala grandiflora* X 7

EUPHORBIACEAE Spurge Family

Important diagnostic features of this family are the **curved or slanting, fine striations** in the seed wall as seen in broken or cut sections; a **straight line** (often slightly grooved) extending from end to the other; and the usual presence of a **terminal knobby outgrowth, the caruncle**. Seed shapes are various: compressed-ovoid to compressed-ellipsoid, or spheroid, quadrangular, or flattish, like thin slices of cake. Cotyledons spatulate, thin, often with a cavity between them.

CROTON Doveweed (pls. 174–178)

Usually **compressed, and ovoid to ellipsoid** or oblong, but some species spheroid; 3–5 mm. long; generally the flattish side consists of two gently sloping faces, the junction of which is marked by a **straight, narrow line** extending from one end to the other; the end bearing the caruncle is sloping-truncate on one side; surface shallowly reticulate-pitted or smooth; commonly mottled with grayish, black, or brown.

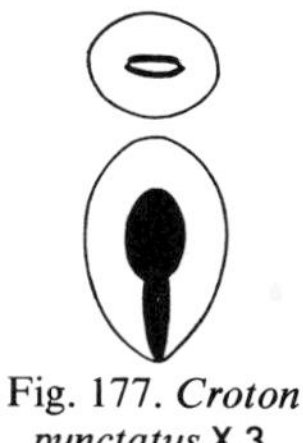

Fig. 177. *Croton punctatus* X 3

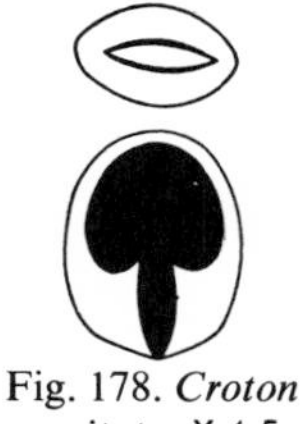

Fig. 178. *Croton capitatus* X 4.5

EREMOCARPUS Turkeymullein (pl. 179)

This is, in effect, a large (4–6 X 3.5 X 2.5 mm.) *Croton*, and was formerly classes as such. Its **plane faces are pitched at a steeper angle than in Croton,** and **it has distinctive, large splotches of black** on a brownish or grayish background.

Fig. 179. *Eremocarpus setigerus* X 4.5

ACALYPHA Copperleaf (pls. 180–182)

Ovoid; finely reticulate; smooth, or warty-rough (*A. ostryaefolia*); **small,** 1–2 mm. long; brown, gray, blackish, with or without mottlings; longitudinal line evident.

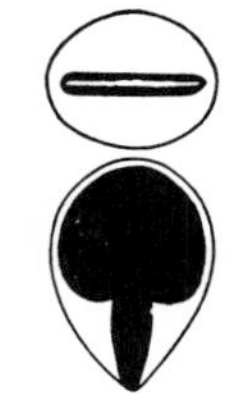

Fig. 180. *Acalypha virginica* X 12

TRAGIA Noseburn (pl. 183)

Globose; 4–5 mm. across; **mottled** brown or grayish; raphe **line short** (2 mm. or less).

Fig. 181. *Tragia urens* X 4.5

STILLINGIA Queensdelight (pls. 184, 653)

Subglobose-ovoid, generally with a **sloping face** near the caruncle, and with the **opposite end broadly truncate**; surface warty-rough to smooth; usually **silvery gray**; 5–7 mm. long.

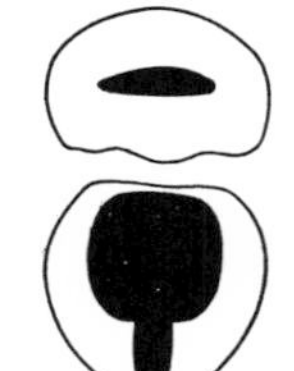

Fig. 182. *Stillingia aquatica* X 4.5

EUPHORBIA Spurge (pls. 185–190)

Rather diverse, but **mainly ovoid, quadrangular-ovoid, or ovate plano-convex, with a distinct longitudinal line;** smooth or warty-rough; white, cream, brown, or black; 2–5 mm. long, except in the subgenus *Chamaesyce*, which is distinct in having seeds 1 mm. or less long, and quadrangular (ovoid-quadrangular, oblong-quadrangular, or ovoid-triangular), commonly with a few, low cross ridges.[17] Embryo variations are indicated in figures 183–185.

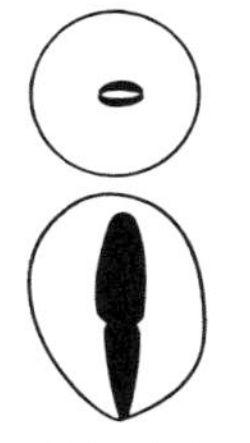

Fig. 183. *Euphorbia lathyrus* X 3

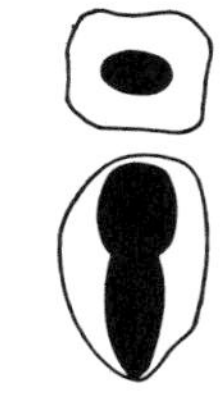

Fig. 184. *Euphorbia vermiculata* X 15

Fig. 185. *Euphorbia dentata* X 7

EMPETRACEAE Crowberry Family

EMPETRUM Crowberry (pl. 739)

Flattish and rounded-triangular, **like thin sectors of a sphere;** narrow edge nearly straight, except for a low, **umbilicus-like protrusion**; woody, finely rough-surfaced; brown; 2 X 1.2 X 0.5 mm.

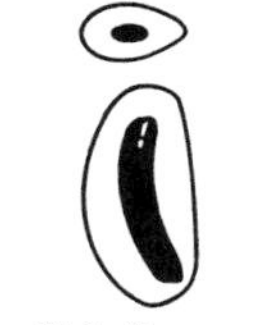

Fig. 186. *Empetrum nigrum* X 10

ANACARDIACEAE Cashew Family

Seeds woody or bony; compressed to flattish and oval, ovate, or elliptic, or ovoid to globular; endosperm plentiful to scant.

RHUS Sumac (pls. 741–744)

Elliptic-ovoid, or somewhat compressed to flattish, and oval or nearly circular; 2–7 mm. long; wall with **2 striate layers and a thin layer between**; smooth; brown or gray.

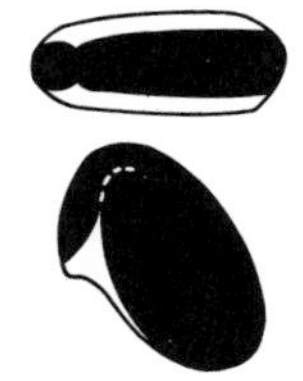

Fig. 187. *Rhus glabra* X 6.5

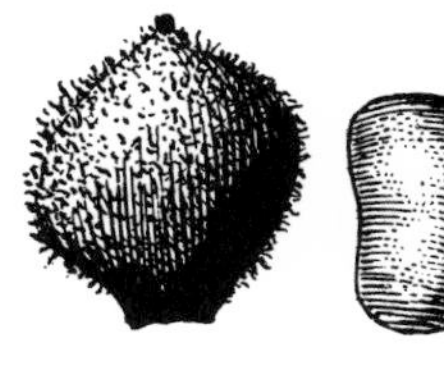

Fig. 188. *Rhus copallina*

[17]See the study *Euphorbia*, Subgenus *Chamaesyce*, in Canada and the United States Exclusive of Southern Florida, by Louis Cutter Wheeler, Rhodora 43:97–154, 168–205, 223–286, 1941.

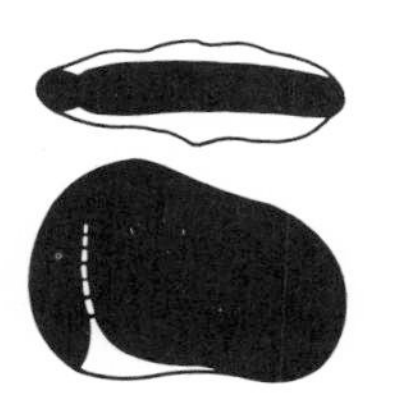

Fig. 189. *Toxicodendron diversiloba* X 5.5

TOXICODENDRON Poison-ivy (pl. 740)

Compressed, ovate to oblong; with lengthwise **irregular ridges and grooves**; wall with **1 thick, striate layer**; about 4 mm. long; light brown.

AQUIFOLIACEAE Holly Family

ILEX Holly (pls. 745–749)

Holly seeds (nutlets) vary in shape from flat and ovate or elliptic to rounded-triangular (like sectors of a sphere), and range in length from 3 to 6 mm.; irregularly ridged lengthwise, or smooth; wall woody; with **fibers running crosswise,** as evident on surface of smooth species or on edges of cut or broken rough-surfaced ones.

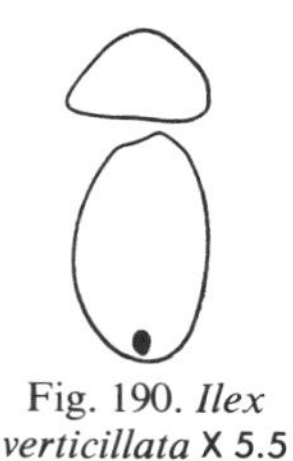

Fig. 190. *Ilex verticillata* X 5.5

CELASTRACEAE Bittersweet Family

Seeds **enclosed** partly or entirely **in an aril**; generally ovoid to ellipsoid; cotyledons spatulate, thin, in fleshy endosperm.

CELASTRUS Bittersweet (pl. 750)
EUONYMUS Wahoo (pl. 751)

Seeds of these two genera are similar in size (generally 3–5 mm. long), shape (ovoid to ellipsoid), and color (**reddish brown**). The surface is marked by a **lengthwise line, ridge, or groove**. Common bittersweet, *C. scandens,* is distinct in being larger (5–6 mm. long), more regularly ellipsoid, and slightly arched, though one side is nearly straight.

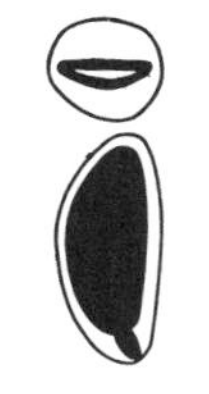

Fig. 191. *Celastrus scandens* X 2.5

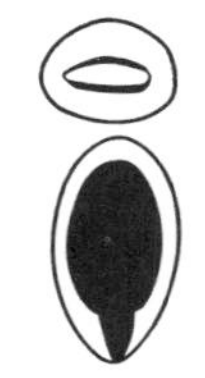

Fig. 192. *Euonymus atropurpureus* X 2

ACERACEAE Maple Family

ACER Maple (pls. 752–755)

The **winged seeds** (samaras) of different species can generally be distinguished by breadth and length of the wings, by shape and size of the body, and by the angle of attachment of the paired seeds.[18] Embryo bent or folded, sometimes greenish; endosperm none.

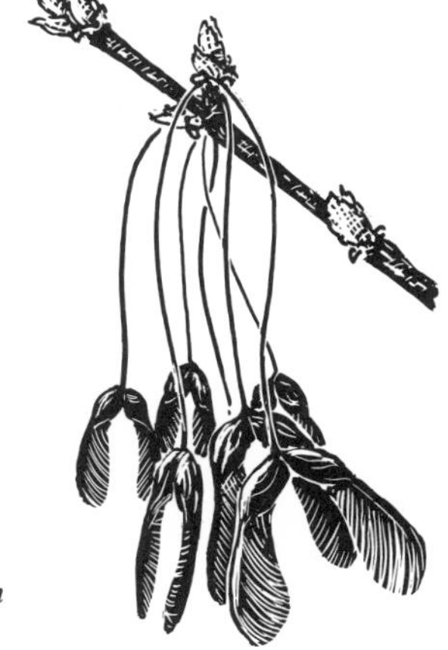

Fig. 193. *Acer rubrum*

BALSAMINACEAE Balsam Family

IMPATIENS Jewelweed (pl. 816)

Quadrangular-ellipsoid, narrowed toward base, with **length-wise ridges;** surface commonly warty or wrinkled irregularly; purplish brown or reddish brown; about 4 X 2.5 X 2 mm.; **embryo greenish, bluish, or yellow.** The two common species cannot be distinguished readily by their seeds.

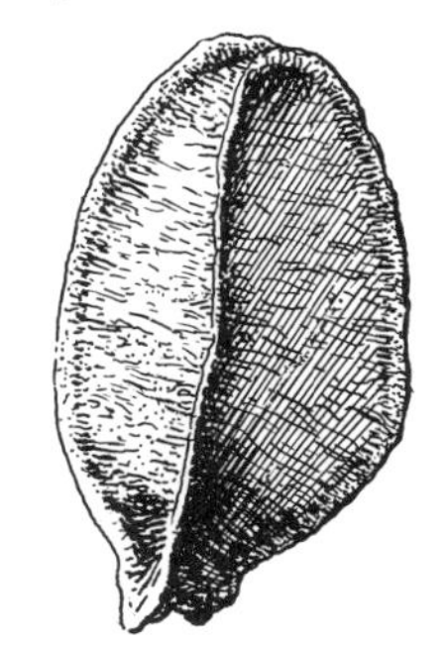

Fig. 194. *Impatiens biflora*

RHAMNACEAE Buckthorn Family

Seeds (some of them nutlets) diverse in size and shape; many broadly rounded-triangular, plano-convex, or globose; wall hard, bony or woody; cotyledons spatulate; endosperm generally present.

RHAMNUS Buckthorn (pls. 756, 757)

Shape varies considerably (figs. 195–197, cross sections), **mainly rounded-triangular** (*R. cathartica* acutely rounded-triangular) or **plano-convex,** usually with a **bony attachment base** suggestive of a tick head; certain species such as *R. crocea* and *ilicifolia* have

[18]For these and other characteristics, see Woody-Plant Seed Manual, U. S. Dept. Agr. Misc. Publ. 654, 1948.

a wedge-shaped opening or groove on one face; plano-convex forms have a distinct lengthwise line on the flat face; 5–10 mm. long; black, brownish, or olive, dull.

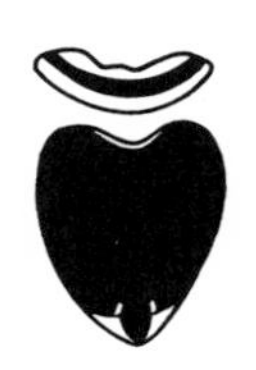

Fig. 195. *Rhamnus alnifolia* X 4.5

Fig. 196. *Rhamnus cathartica* X 3.5

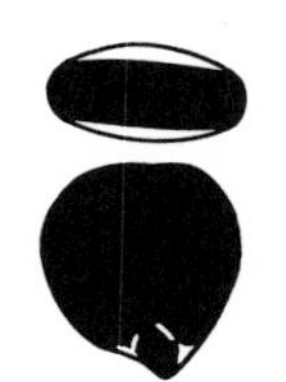

Fig. 197. *Rhamnus purshiana* X 3.5

CEANOTHUS Buckbrush, Jerseytea (pls. 760, 761)

Ovoid rounded-triangular, with a broadly rounded back and two nearly plane faces joined in a low, obscure ridge; **both ends broadly rounded;** 2–4 mm. long; black, brown, or olive, dull or glossy.

Fig. 198. *Ceanothus cuneatus* X 5

CONDALIA Condalia (pl. 759)

Circular plano-convex; 2–4 mm. across; brown, dull; borne in pairs in 2-celled, woody, broad compressed-ovoid nutlets.

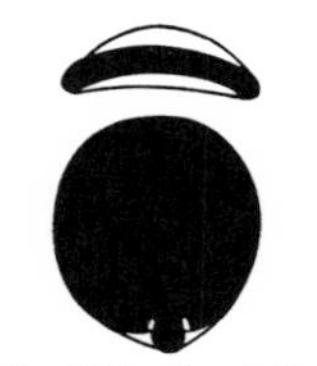

Fig. 199. *Condalia obovata* X 4.5

BERCHEMIA Supplejack (pl. 758)

Oblong-cylindric, slightly compressed, with **one end open** and a **shallow groove** extending over both sides, because of the **2 cells inside**; woody; light brown; 4 X 3 X 2 mm.

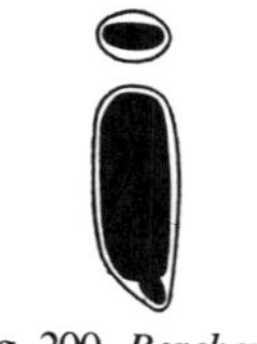

Fig. 200. *Berchemia scandens* X 4

VITACEAE Grape Family

Seeds woody; **wall folded inward in deep grooves,** strongly striate in section; generally **ovoid rounded-triangular**, but some are plano-convex or subglobose; embryo small, basal, in firm-fleshy endosperm.

VITIS Grape (pls. 762–764)

Ovoid rounded-triangular (sometime elongate), plano-convex (*V. rotundifolia*), or simply ovoid; marked on the roundish face by a **central oval area connected by a groove over the broad end** of the seed, and on the plane faces by **broad, shallow grooves,** on both sides of the central ridge; brown to black; 4–7 mm. long.

Fig. 201. *Vitis aestivalis* X 4

AMPELOPSIS Peppervine (pl. 765)
PARTHENOCISSUS Woodbine (pl. 766)

Both are similar to *Vitis*, but proportionately broader and shorter. Woodbine seeds are distinct in the **flatness of their plane faces**, in the sharply defined margins, and in the **comparative thinness of the walls.** *Ampelopsis* is **more rounded** (subglobose in *A. cordata*) and **heavy-walled.**

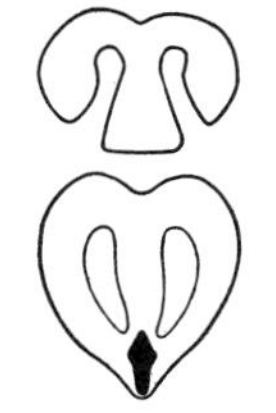

Fig. 202. *Ampelopsis cordata* X 3.5

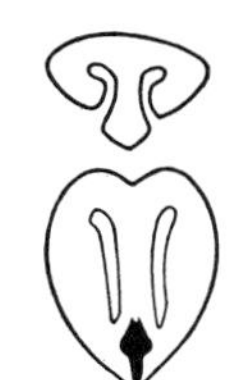

Fig. 203. *Parthenocissus quinquefolia* X 4

MALVACEAE Mallow Family

Seeds **compressed-reniform** to flat or globose; often with external contours showing that the embryo is bent or arched; **cotyledons large, thin, and folded,** endosperm usually present.

SPHAERALCEA Globemallow (pl. 191)
SIDALCEA Checkermallow (pl. 192)
MALVASTRUM Falsemallow (pl. 193)

These closely related genera are difficult to distinguish by their seeds because of overlapping characteristics. All have a **deep, broad notch** in the **attachment area** and are **compressed-reniform, with one end narrowed**; many are coated with hairs or minute protuberances. *Sidalcea* is distinct in being **considerably thicker along its arched back,** sloping down to a comparatively thin inner margin, whereas *Sphaeralcea* and *Malvastrum* are more uniformly compressed or flattish. *Sphaeralcea* generally is obliquely reniform and has a coating of hairs, whereas *Sidalcea* is smooth and *Malvastrum* generally so.

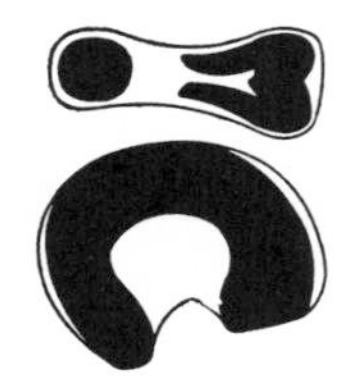

Fig. 204. *Sphaeralcea angustifolia* X 12

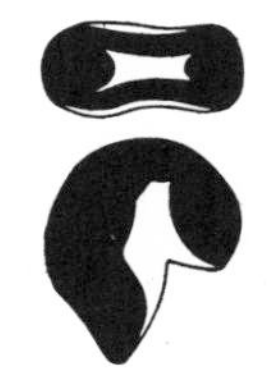

Fig. 205. *Sidalcea neomexicana* X 7.5

Fig. 206. *Malvastrum exile* X 15

SIDA Sida (pl. 194)

Rounded-triangular, with two flat faces and a rounded back, like sectors of a sphere, the inner edge narrow and the outer edge of variable breadth in different species; the **truncate top** with a **conspicuous notch beside a pommel-like knob;** about 2 mm. long; dark brown to black.

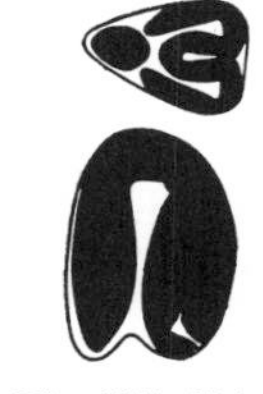

Fig. 207. *Sida rhombifolia* X 9

HIBISCUS Hibiscus, Rosemallow (pls. 654, 655)

Subglobose or ovoid, to compressed or flattish, reniform, generally with a **conspicuous attachment scar beside a saddle-pommel knob;** surface in most species **marked by minute, scurfy protuberances,** in rows, or coated with hairs (*H. militaris, syriacus, coulteri, denudatus*), or smooth (*H. esculentus, furcellatus*); 2–5 mm. long; brown, olive-green to black.

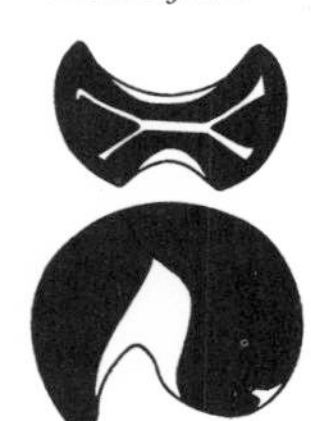

Fig. 208. *Hibiscus incanus* X 7.5

STERCULIACEAE Chocolate Family

MELOCHIA Chocolateweed

Rounded-triangular, pointed near one end and rounded at the other (no notch); top capped by brownish, papery material, the removal of which shows a circular area marked by **fine striations radiating from the center**; dark gray or dark brown, splotched crosswise with small, dark markings; 3 X 1.5 X 1.5 mm.

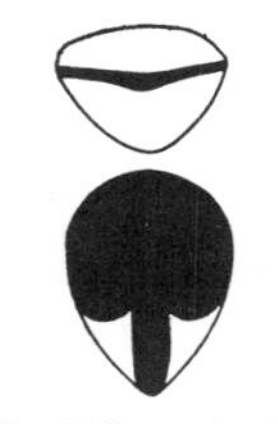

Fig. 209. *Melochia corchorifolia* X 7.5

HYPERICACEAE St. Johnswort Family

Seeds minute (0.5–2 mm. long); rodlike to oblong-cylindric or ellipsoid, like some insect droppings; finely to coarsely **cellular-reticulate** and marked by a **lengthwise line** and by a nipple-like point at one end.

HYPERICUM St. Johnswort (pl. 817)
ASCYRUM St. Andrewscross (pl. 818)

These two genera are similar in the family characteristics noted above. *Ascyrum* seeds, however, are slightly thicker, average 0.75–1 mm. long, and are uniformly dark (black or dark brown). *Hypericum* seeds vary more in length and color: some slender, slightly arched, black ones (*H. nudiflorum, prolificum,* etc.) are about 2 mm. long; a few cream-colored ones are less than 1 mm. long (*H. gentianoides, punctatum*); and still others are brown and intermediate in length (*H. canadense, denticulatum, virginicum,* etc.).

VIOLACEAE Violet Family

VIOLA Violet (pls. 195–197)

Ovoïd, with a lengthwise line on one side and scurfy material (the caruncle) at the attachment end; cream, brown, dark olive, mottled, or black; generally about 2 mm. (1.5–3 mm.) long.

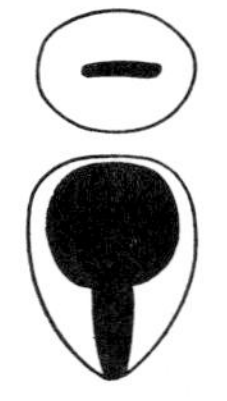

Fig. 210. *Viola eriocarpa* X 7.5

PASSIFLORACEAE Passionflower Family

PASSIFLORA Passionflower (pls. 198, 199)

Flattish, ovate to elliptic; coarsely marked by **shallow pittings**, or by crosswise ridges and narrow lengthwise lines (like **longitude and latitude lines**); 4–6 mm. long; wall thick, woody, striate; cream to black.

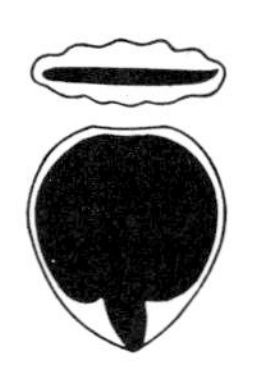

Fig. 211. *Passiflora incarnata* X 3

LOASACEAE Loasa Family

MENTZELIA Prairiestar (pls. 200, 201)

Diverse in shape and size; flat, thin, and with an encircling wing (*M. decapetala, laevicaulis, multiflora, stricta,* and *wrightii*), or 3-sided and truncate, with grooves in each angle (*M. dispersa, micrantha*), or irregularly angled (*M. lindleyi*), or flattish-oblong, with a low ridge on one side (*M. oligosperma*). Whether the seed is winged or not, the surface usually has minute, papillae-like protuberances, which give it a characteristic **beady, reptile-skin** appearance. Length varies from 1 to 4 mm.; and color from cream to dark, spotted with gray.

Fig. 212. *Mentzelia decapetala* X 5.5

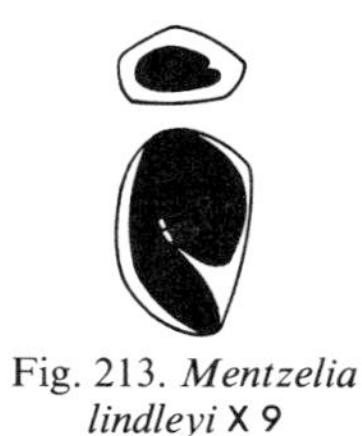

Fig. 213. *Mentzelia lindleyi* X 9

CACTACEAE Cactus Family

Seeds of this family are similar internally in being bent, with little or no endosperm; externally, they are of two main types: (1) comparatively large, flattish, light-colored (*Opuntia*); or (2) smaller and generally dark (other genera).

OPUNTIA Pricklypear (pls. 202, 203)

Flattish-subcircular, either **with a distinct groove parallel to the margin** (as in *O. compressa, engelmannii, erinacea, humifusa, macrocentra, polycantha, pollardi*), **or without a bordering groove** (as in *O. imbricata, leptocaulis, parishii, spinosa*). Cream to light brown; 3–6 mm. across..

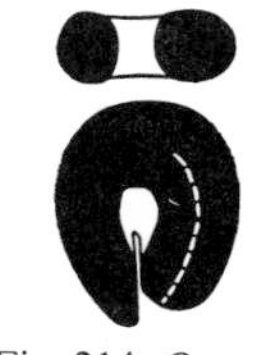

Fig. 214. *Opuntia leptocaulis* X 5.5

Fig. 215. *Opuntia pollardi* X 4.5

CEREUS Giantcactus (pl. 204)

Compressed-ovoid, with a **truncate, hollow base;** black, glossy, and obscurely cellular-reticulate (*C. giganteus, pentagoneus*), or dull and rough or pitted (*C. greggii*); 2–4 mm. long; endosperm none.

MAMMILLARIA Cactus (pl. 205)

Diverse in shape and surface characteristics: ovoid (*M. missouriensis*) to compressed-ovoid (*M. micromeris, vivipara*), or flattish-reniform (*M. robustispina*). *M. missouriensis* is black and rugose-pitted and about 1.5 mm. long; *M. micromeris* is black and nearly smooth, about 2 mm. long, and has a large cavity; *M. robustispina* is brown, smooth, cellular-reticulate, and 4 mm. long; *M. vivipara* is also brown, but is cellular-pitted, and 2 mm. long.

Fig. 216. *Mammillaria robustispina* X 5

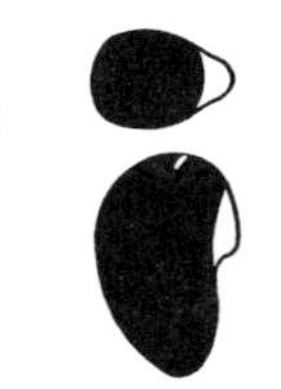

Fig. 217. *Mammillaria vivipara* X 10

ECHINOCACTUS Cactus (pl. 206)

Ovoid to compressed-oval or somewhat reniform; all with a **cavity** at or near the base and usually **black**; most species with surface finely or coarsely roughened by protuberances or pittings, but a few glossy (*E. texensis, xeranthemoides*); 1–4 mm. long.

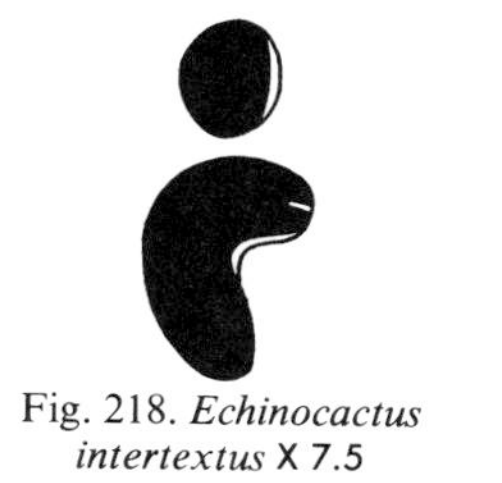

Fig. 218. *Echinocactus intertextus* X 7.5

Fig. 219. *Echinocactus polycephalus* X 5

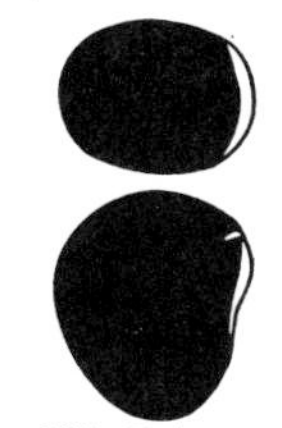

Fig. 220. *Echinocactus setispinus* X 14

ELAEAGNACEAE Elaeagnus Family

Seed enclosed in a hard, persistent covering (partly calyx) in *Elaeagnus*; in *Shepherdia*, borne in a berry-like fruit from which it escapes readily.

ELAEAGNUS Russianolive (pl. 767)

Ellipsoid, pointed-ellipsoid, or spindle-shaped, with **dark and light lengthwise ribs alternating** on the hard, fibrous wall; light stripes commonly silvery whitish; 6–10 mm. long. Endosperm none.

SHEPHERDIA Buffaloberry (pl. 768)

Compressed-oval, some plano-convex; a notch on one side of base and a lengthwise groove from notch upward on both faces; glossy, black or dark brown; 4–5 mm. long.

Fig. 221. *Shepherdia argentea* X 5.5

LYTHRACEAE Loosestrife Family

DECODON Waterwillow (pl. 656)

Prismatic-wedge-shaped or pyramidal, with **2, 3, or more plane faces, narrowed downward from a broad, slightly rounded top to a pointed base;** varnished brown with darker lines; 1.5 X 1 X 1 mm. Endosperm none.

ONAGRACEAE Evening-primrose Family

Seeds minute, small, or medium; shapes various, often elliptic; endosperm none.

JUSSIAEA Waterprimrose (pls. 657, 658)

In certain species (*J. leptocarpa, pilosa, suffruticosa*) the seeds are partially **enclosed in a heart-shaped block of corky material**, and in *J. diffusa* they are buried in oblong-truncate, oblique chunks of corky wood. The true seeds are **minute**, less than 1 mm. long; **ellipsoid or arched-ellipsoid,** with narrowed or obtuse ends; one side is somewhat compressed and has a thin, dark, lengthwise flange; cream to brown.

OENOTHERA Evening-primrose (pls. 207-209)

Diverse; **variously angled,** often with a rounded side, or angular-ellipsoid, spindle-shaped to flattish-oblong, or angular-lanceolate; smooth- to rough-surfaced, thin- to thick-walled; yellow-brown to dark purplish brown; 1–4 mm. long.

GAURA Gaura (pl. 210)

Seeds (nutlike fruits) woody; **generally 4-sided, spindle-shaped,** except in *G. parviflora*, which is elongate-ellipsoid with irregular rounded ridges and grooves; in most species, the seeds are hairy and have a **lengthwise rib median between the angles** and extending halfway up from the base; about 6 mm. long.

TRAPACEAE Waterchestnut Family

TRAPA Waterchestnut, Watercaltrop (pl. 555)

The hard, large (3–4 cm. across), **nutlike seeds** (mericarps) are somewhat top-shaped and are armed on their thick median zone with **4 stout, retrorsely barbed spines,** which are attached to bulbous bases and make the seed somewhat 4-angled; dark to black. Endosperm none.

HALORAGIDACEAE Watermilfoil Family

Seeds (nutlets) woody, oblong or oblong rounded-triangular to ellipsoid, or triangular; embryo linear in endosperm of variable amounts.

HIPPURIS Marestail (pl. 556)

Oblong-cylindric, with short points on both of the obtuse ends; an obscure line extends from top to bottom, on both sides of seed; wall thick, soft-woody; embryo circular in cross section; 2 X 1 X 1 mm.

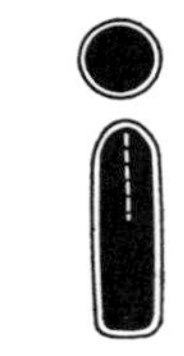

Fig. 222. *Hippuris vulgaris* X 10

MYRIOPHYLLUM Watermilfoil (pls. 557, 558)

Oblong rounded-triangular, 1.5–2 mm. long, except in *M. tenellum*, which is oblong-cylindric and barely 1 mm. long. *M. scabratum* has its outer (rounded) margins fringed with toothlike processes; *M. spicatum* is obscurely warty on its back (rounded) surface. Worn seeds generally are open at one end.

Fig. 223. *Myriophyllum spicatum* X 11

PROSERPINACA Mermaidweed (pl. 559)

Broadly triangular (ovoid-pyramidal), with thin margins in *P. palustris* and rounded margins in *P. pectinata*; reddish brown; 4 X 3 X 3 mm.; interior distinctive in showing **3 cylindric cells**, one in each of the angles.

Fig. 224. *Proserpinaca pectinata* X 7.5

ARALIACEAE Aralia Family

ARALIA Aralia (pls. 770, 771)

Flattish-oblong, pointed near base; two distinct groups of species: (1) seeds small (about 2 mm. long), **whitish with a satin sheen,** wall rather thin and with crosswise surface fibers (*A. californica, cordata, racemosa*); (2) seeds larger (2.5–4 mm. long), **thick-walled, rough-surfaced, brown** (*A. hispida, nudicaulis, spinosa*).

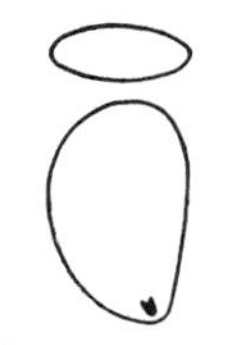

Fig. 225. *Aralia hispida* X 7

UMBELLIFERAE Parsley Family

Seeds (mericarps) usually flattish, plano-convex, or concave-convex; outlines various: circular, oval, or elliptic, to narrow and elongate. Surface has **several lengthwise ribs or ridges**; embryo small, basal in fleshy firm endosperm.

CENTELLA Centella (pl. 659)
HYDROCOTYLE Pennywort (pl. 660)

Flat, with inner edge straight, outer edge rounded; often a pair of seeds remaining united by their inner edges and appearing **kidney-shaped. Centella** is distinct in having 2 or 3 branched, prominent, curved, **veinlike ribs,** whereas *Hydrocotyle* has, if any, 1 or 2 straight, less prominent, and **unbranched ribs**. *Centella* seeds are 4 X 3 X 0.5 mm.; *Hydrocotyle* seeds are slightly smaller.

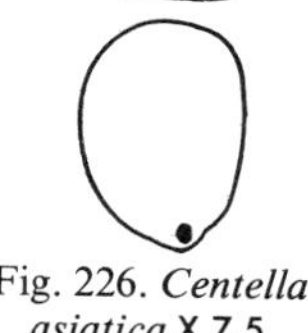

Fig. 226. *Centella asiatica* X 7.5

Fig. 227. *Hydrocotyle umbellata* X 12

DAUCUS Wildcarrot, Queen-Annes-lace (pl. 211)

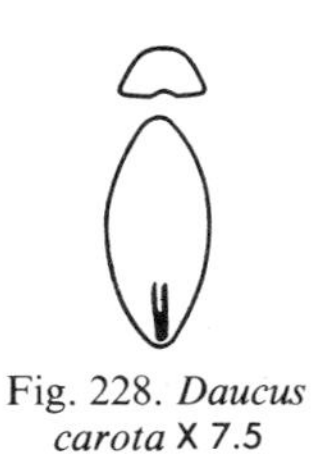
Fig. 228. *Daucus carota* X 7.5

Elliptic plano-convex, with the margins and 3 ridges on the back bearing **long, narrow, whitish, flat, spinelike processes hooked at the end** and with bristly hairs in rows between the crowned ridges; 3–4 mm. long.

CORNACEAE Dogwood Family

Seeds (pits of drupes) **hard, bony-woody**; globose, ellipsoid, or compressed-oval to compressed elliptic; embryo spatulate in fleshy endosperm.

CORNUS Dogwood (pls. 772–775)

Seed generally **globose to ellipsoid;** containing **2 seed chambers,** or often only one; **lengthwise groove** or line all the way around, from top to bottom. The ellipsoid forms (*C. canadensis, florida, nuttallii, sessilis*) have no ridges, except for low obscure ones in *C. sessilis,* whereas ridges are usually evident on the globose forms.

Fig. 229. *Cornus canadensis* X 7

Fig. 230. *Cornus amomum* X 3.5

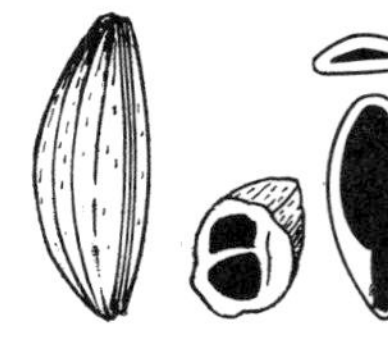
Fig. 231. *Cornus florida.* Diagram (right) X 2.5

NYSSA Blackgum (pl. 769)

Fig. 232. *Nyssa biflora* X 2

The two widespread species, *N. biflora* and *sylvatica*, are similar in being compressed-ovoid or compressed-ellipsoid, and in having several **grooves or stripes** on each face; about 7 X 5 X 3 mm.; *N. aquatica* and *ogeche* are much larger (about 2 cm. long) and are rough, with deep grooves.

ERICACEAE Heath Family

Seeds (some of them nutlets of drupes) **diverse**; minute and delicate; or medium small and compressed, oval to elliptic; with a hard, glossy, often reticulate surface; or irregular, bony nutlets. Embryo linear to spatulate in fleshy endosperm.

GAULTHERIA Wintergreen (pl. 776)

Flattish, irregularly angled and curved, often ovate or wedge-shaped, with a broad, rounded top; glossy brown; about 1 mm. across; the cellular reticulations are obscure on *G. procumbens* and distinct on *G. shallon* (which is less glossy).

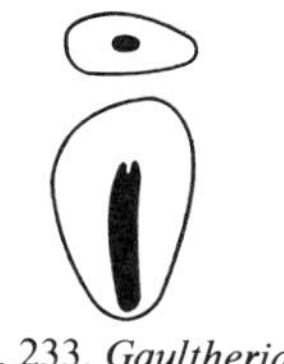

Fig. 233. *Gaultheria shallon* X 17

ARCTOSTAPHYLOS Manzanita, Bearberry (pls. 777, 778)

Rounded-triangular, globose-angular, or globose, the shape often depending upon the number of nutlets adhering to each other; about 5 mm. long; **walls heavy and dark brown.** In *A. diversifolia*, all the nutlets remain cemented to each other in a solid ball; in *A. pungens* they separate individually; and in others 2 or 3 nutlets generally cling to each other.

ARBUTUS Madrone (pl. 779)

Compressed, oval to elliptic, or irregularly angled, **often concave-convex** and incurved-pointed at ends; dark brown, dull, with very narrow, **linear cellular reticulations**; 3 X 1.5 X 1 mm.; wall rather thin.

Fig. 234. *Arctostaphylos uva-ursi* X 6

Fig. 235. *Arctostaphylos andersonii* X 5

Fig. 236. *Arctostaphylos bicolor* X 5

Fig. 237. *Arbutus menziesii* X 6

GAYLUSSACIA Huckleberry, Dangleberry (pls. 780, 781)

Flat or flattish, oval and somewhat wedge-shaped, with one edge thin; surface minutely rough, somewhat **like sandpaper**; light brown; about 2 mm. long; embryo linear.

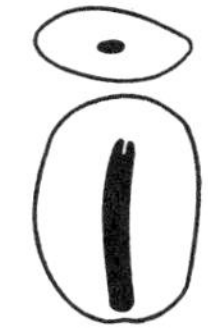

Fig. 238. *Gaylussacia frondosa* X 9

VACCINIUM Blueberry, Deerberry (pls. 782–784)

Diverse; from variously angled to compressed-elliptic or compressed-oval to ellipsoid; cellular-reticulate; brown to dark brown; 1–2 mm. long; embryo linear.

Fig. 239. *Vaccinium angustifolium* X 13

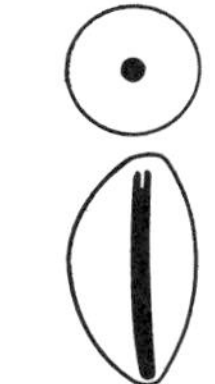

Fig. 240. *Vaccinium macrocarpon* X 9

SAPOTACEAE Sapodilla Family

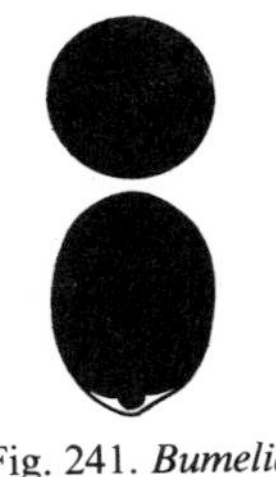

Fig. 241. *Bumelia lanuginosa* X 3

BUMELIA Bumelia (pl. 785)

Ellipsoid to **ovoid or globose**; a cavity at the smaller end, except in *B. lycioides*, which has a circular scar; **shiny, brown;** 5–12 mm. long; wall hard, brittle, veined on the interior.

EBENACEAE Ebony Family

Fig. 242. *Diospyros virginiana* X 1

DIOSPYROS Persimmon (pl. 786)

Flat-elliptic (*D. virginiana, kaki*), or compressed-ovoid to plano-convex (*D. texana, lotus*); often with one edge nearly straight; cellular lines diagonal on *D. virginiana* and *texana*, irregularly wavy and obscure in Asiatic species; brown, about 1.5 cm. long in *D. virginiana*; black, 1 cm. long in *D. texana*; wall and endosperm very hard.

OLEACEAE Olive Family

Seeds diverse, because some are in samaras or keys (as in *Fraxinus*), some are in nutlets of drupes (as in *Forestiera*), or in a berry (as in *Ligustrum*), or are borne in capsules (other genera); embryo generally spatulate in fleshy endosperm.

FRAXINUS Ash (pls. 788–789)

Species of ash can be distinguished to some extent by the samaras, particularly as regards their size and shape and the extension of the wing along the seed body. The true seeds inside the samaras are linear-cylindric to flat-oblong, are about 1 cm. long, and generally show **lengthwise wavy lines in grooves**; the embryo as viewed in cross section is sandwiched between layers of endosperm.

Fig. 243. *Fraxinus quadrangulata* X 1

Fig. 244. *Fraxinus excelsior* X 1

Fig. 245. *Fraxinus oregona* X 1.5

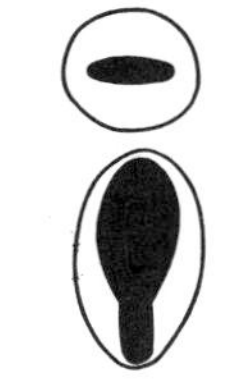

Fig. 246. *Forestiera angustifolia* X 4

FORESTIERA Waterprivet (pl. 787)

Seeds (nutlets) narrowly elliptic (as in *F. acuminata*) or **broadly elliptic to ovoid** (as in other species); **pointed near the ends** and often arched slightly, with numerous **branched, lengthwise veins** in most species (lengthwise ridges and grooves in *F. acuminata*); about 1.5–5 mm. long.

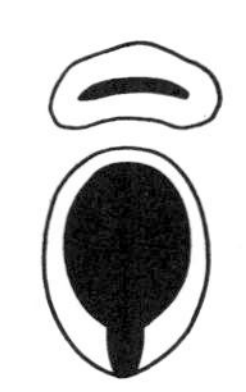

Fig. 247. *Ligustrum vulgare* X 4

LIGUSTRUM Privet (pl. 790)

Rather **irregular in shape and surface;** generally oval to elliptic in outline, and **plano-convex, compressed,** or ellipsoid; surface dull, rather scurfy and wrinkled; mainly 2–4 mm. long.

LOGANIACEAE Logania Family

Fig. 248. *Gelsemium sempervirens* X 4

GELSEMIUM Jessamine (pl. 791)

Flat, elliptic, winged; 9 X 4 X 0.5 mm.; wing very thin, irregularly margined toward end; seed-body area also thin, not definitely bounded, except by many conic-oblique papillae and by darker brown color.

ASCLEPIADACEAE Milkweed Family

Fig. 249. *Asclepias humistrata* X 3

ASCLEPIAS Milkweed (pl. 212)

Flat, generally margined by an encircling wing, obovoid and with a terminal tuft of silky hairs (except in *A. perennis*); seed area slightly thicker and clearly differentiated; brown to reddish brown; 5–10 mm. long.

CONVOLVULACEAE Morning-glory Family

Shape mainly rounded-triangular (*Convolvulus, Ipomoea, Jacquemontia*), though in some genera globose (all *Cuscuta* and certain *Ipomoea* species), ovoid-truncate (*Breweria*), or other shapes; surface dull, granular-rough or hairy; embryo bent (coiled, in *Cuscuta*), **cotyledons folded,** except in *Cuscuta*; **endosperm hard, semitransparent.**

Fig. 250. *Breweria pickeringii* X 3.5

BREWERIA   Breweria (pl. 213)

Ovoid-truncate, with a V-shaped notch and a rounded hump at the truncate attachment end; dull, minutely rough, brown; 3–4 X 2.5 X 2.5 mm.

Fig. 251. *Convolvulus incanus* X 5

CONVOLVULUS   Morning-glory (pls. 214, 215)
IPOMOEA   Morning-glory (pls. 217–219)

These two genera are similar in having their seeds mainly rounded-triangular. However, in *Convolvulus* the **attachment scar is nearly at right angles** to the seed's long axis, whereas in *Ipomoea* (except in large-seeded species such as *I. bona-nox, dissecta, pres-caprae, tuba*) **the scar is usually almost parallel** to the long axis. *Convolvulus* seeds average smaller, 3–4 mm. long, are dull, minutely warty in *C. arvensis* and *tricolor*, and black to brown; *Ipomoea* seeds are more diverse: lanceolate-ovoid (*I. quamoclit*) to globose, black to brown, dull to hairy (*I. fistulosa, longifolia, pandurata, pres-caprae, sagittata*), and 4–12 mm. in length.

Fig. 252. *Ipomoea lacunosa* X 3.5

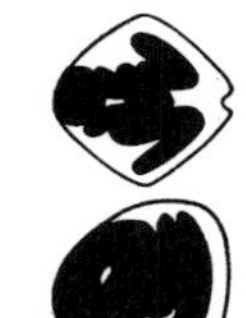

Fig. 253. *Jacquemontia tamnifolia* X 6

JACQUEMONTIA   Jacquemontia (pl. 216)

Rounded-triangular and truncate; **minutely warty;** brown; comparatively small, 2 X 1.5 X 1.5 mm.

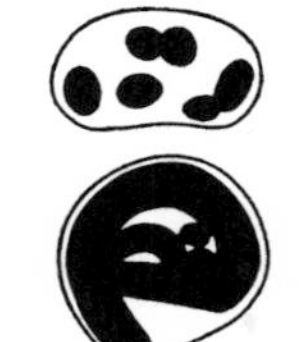

Fig. 254. *Cuscuta pentagona* X 15

CUSCUTA   Dodder (pl. 220)

Small, about 1 mm. across; globose to globular or rounded-triangular; dull or scurfy, brown; **embryo yellowish, coiled,** in semi-transparent, hard endosperm.

POLEMONIACEAE   Phlox Family

Seeds generally either flattish plano-convex (*Collomia, Phlox, Microsteris*) or irregularly angled (*Gilia, Navarretia, Polemonium*); about 2.5 mm. long in the flattish type and generally smaller, 0.5–2 mm. long, in the irregularly angled forms.

GILIA Gilia (pls. 221, 222)

Irregularly angled, with the edges sharp or rounded, except *G. gilioides*, which is concave-ovoid; mainly 0.5–1 mm. long, though *G. aggregata* and *gilioides* are about 2 mm. long; smooth or somewhat wrinkled.

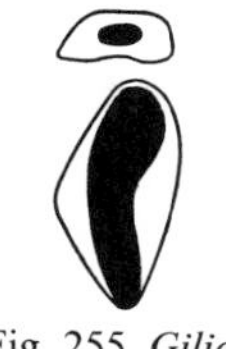

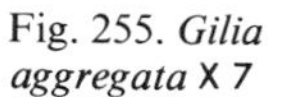

Fig. 255. *Gilia aggregata* X 7

Fig. 256. *Gilia gracilis* X 6

COLLOMIA Collomia (pl. 223)

Elliptic, and plano-convex to flattish or compressed (*C. grandiflora*), or linear and concave-convex (*C. linearis*); all species with a **lengthwise line or groove** on one face; brown, dull; 2.5–3 mm. long.

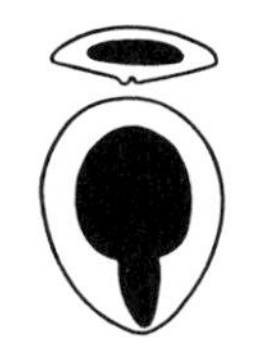

Fig. 257. *Collomia grandiflora* X 5.5

PHLOX Phlox (pl. 224)

Elliptic and flattish-compressed or flattish plano-convex; with a **lengthwise groove** on one face; surface coarsely or finely rugose; dark brown to black; 2–3.5 mm. long.

Fig. 258. *Phlox pilosa* X 5

BORAGINACEAE Borage Family

Seeds (nutlets) **ovoid, ovoid-angular**, compressed-ovate, or other shapes; many with **surface hard** or bony and glossy (*Lithospermum, Onosmodium, Myosotis, Amsinckia, Plagiobothrys*), but many others dull, thin- to heavy-walled and some with hooked hairs; endosperm scant or none.

HELIOTROPIUM Heliotrope (pl. 661)

Rather **diverse; globose to ovoid, concave-convex, or plano-convex;** 1–3 mm. long; edges rounded (*H. curassavicum* and most others), or sharp-edged, acute-beaked, and with lengthwise ridges in *H. indicum*; hairy on the roundish back (*H. angustifolium, greggii, inundatum, parviflorum, phyllostachyum, tenellum*), or smooth (*convolvulaceum, curassavicum, europaeum, indicum, parviflorum,* and *xerophyllum*). Two **eyelike areas** on the concave or plane face distinguish *H. inundatum* and *phyllostachyum*.

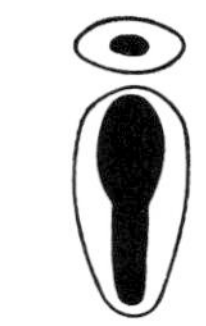

Fig. 259. *Heliotropium curassavicum* X 10

LITHOSPERMUM Gromwell (pls. 228, 229)

Ovoid-truncate, with a ridge on one side, and a broad, flattish scar at the base; surface smooth, glossy in most species (coarsely rough in *L. arvense, incisum*); wall hard and crusty; 2–4 mm. long.

MYOSOTIS Forget-me-not (pl. 230)

Compressed-flattish and ovate, with thin margins; dark **brown to black and glossy,** but **attachment scar whitish;** slightly more than 1 mm. long in *M. laxa* and *virginica,* and 1–2 mm. long in *M. scorpioides.*

ECHIUM Blueweed (pl. 231)

Similar to *Lithospermum arvense* in shape and surface; ovoid-truncate, with a **large basal scar** partly surrounded by a low wall or **collar**; surface rough; one side straight and with a low ridge.

AMSINCKIA Fiddleneck (pls. 225–227)

Ovate rounded-quadrangular and arched; with a linear whitish scar on the concave side and a ridge on the roundish back; **rough-warty or cross-wrinkled;** wall hard, brittle; 2–3 mm. long.

PLAGIOBOTHRYS Popcornflower (pls. 232, 233)

Compressed, pointed-ovate, and slightly arched; the arched back has a **low, lengthwise ridge and several crosswise ridges,** while the opposite, slightly concave side has a **puckered-up attachment scar,** except in *P. jonesii,* which has a Y- or T-shaped mark on the back and small, warty pustules over much of the surface. *P. shastensis* and *tenellus* are somewhat cross-shaped. All species grayish or brownish; wall hard, brittle; 1–3 mm. long.

VERBENACEAE Verbena Family

Seeds (nutlets or schizocarps in *Verbena*) diverse; the two genera included here are flattish and oblong to circular; embryo spatulate; endosperm present or lacking.

VERBENA Verbena (pls. 234–236)

Compressed-oblong and slightly rounded-triangular in cross section, making it somewhat 2-sided; the rounded back bearing **lengthwise ridges** toward lower end and a **network of cross ridges** below; inner face of two planes meeting in a low ridge, the surface generally covered with **whitish papillae;** margin bordered by a narrow flange; attachment scar whitish (in *V. ciliata* the scar is surrounded by a cuplike extension); 1.5–3 mm. long; endosperm lacking.

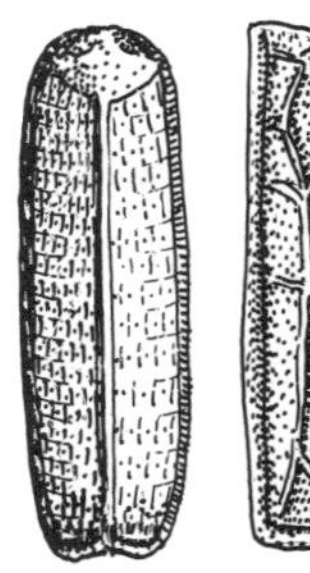

Fig. 260. *Verbena hastata*

CALLICARPA Beautyberry (pl. 792)

Elliptic, flattish or concave-convex, narrowly margined on the concave side and with a central attachment scar; white to light brown; 1.5–2.5 mm. long.

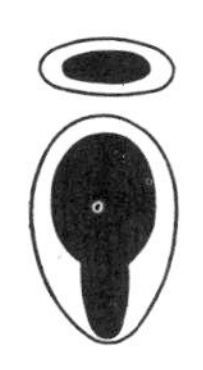

Fig. 261. *Callicarpa americana* X 10

LABIATAE Mint Family

Seeds (nutlets) generally **rounded-triangular** (because they are borne in tight groups of 4, or sometimes 3); surface smooth to ridged or horny; wall thin and brittle to thick and woody; embryo spatulate; endosperm scant or none.

TEUCRIUM Germander (pl. 237)

Globose rounded-triangular, **the back** (outer side) **coarsely reticulate, with low ridges,** and the inner side with a **large, roundish scar** margined by a lighter area; wall thick, woody; 1.5–2.5 mm. long.

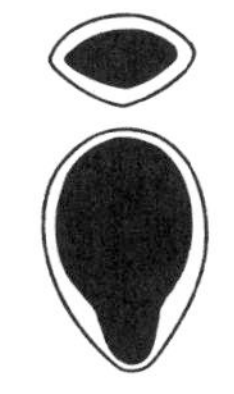

Fig. 262. *Teucrium canadense* X 13

TRICHOSTEMA Bluecurls (pl. 238)

Ovoid to oblong, or globose rounded-triangular, the back (outer side) reticulate-ridged in *T. dichotomum* and *lanatum,* and bearing finger-like projections and pubescence in *T. lanceolatum*; a **large, roundish scar** on the inner side; 1–4 mm. long.

Fig. 263. *Trichostema dichotomum* X 10

SCUTELLARIA Skullcap (pl. 239)

Diverse; globose and covered with dark brown **bracts** or scales in *S. pilosa* and *integrifolia*; other species **circular in outline,** with two faces, one of them more rounded; *S. galericulata* and *lateriflora* are yellowish, with a pebbly-warty surface and a conspicuous "umbilicus" on the inner face; *S. parvula* is like the upper part of a dark **toadstool,** partly covered with small, roundish warts; 1–2 mm. across.

HEDEOMA False Pennyroyal (pl. 241)

Compressed-oblong (*H. ciliata, hispida, nana*) **to subglobose-compressed** (*H. pulegioides*) or rounded-triangular; **minute,** about 1 mm. long; finely cellular-reticulate; most species brownish, but *H. pulegioides* black.

MONARDA Beebalm (pl. 242)

Oblong-elliptic and compressed-triangular, with a **wedge-shaped attachment area** and a small, whitish, mouthlike scar; smooth; 1.5–2 mm. long.

PRUNELLA Selfheal (pl. 240)

Ovoid and compressed-triangular, tipped with white at the base, and the shiny, brown surface **margined with darker brown lines and having a pair of central lengthwise lines;** 2 X 1 X 0.5 mm.

LAMIUM Henbit (pl. 245)

Elongate-ovoid, and rounded-triangular in section; brownish or grayish and **mottled with silvery white;** 2 X 1 X 0.5 mm.

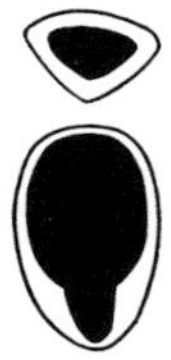

Fig. 264. *Lamium amplexicaule* X 10

SALVIA Sage (pls. 243, 244)

Oblong, elliptic, or ovoid **compressed-triangular,** to globose (*S. officinalis*); surface **generally marbled** with streaks or netted lines; gray, brownish, or black; 2–5 mm. long.

LYCOPUS Bugleweed (pl. 662)

Oblong-ovoid and compressed-triangular, with the top truncate and irregularly notched (*L. americanus, uniflorus, virginicus*), or rounded (*L. europaeus, lucidus, rubellus*); a small, **white mouth at the base** and a marginal flange; brownish with **goldlike globules** on the inner face; 1–1.5 mm. long.

MENTHA Mint (pl. 663)

Globose to ovoid **compressed-triangular, wedge-shaped** at the base; brown, smooth, dull; **minute,** 1 mm. long or less.

SOLANACEAE Nightshade Family

Usually medium small to large, flat or flattish, and oval to circular or reniform, but in genera such as *Nicotiana* and *Petunia* the seeds are minute and globose or subglobose; **embryo coiled or bent in fleshy endosperm.**

SOLANUM Nightshade (pls. 246–249)
PHYSALIS Groundcherry (pls. 250, 251)

Flat, and oval or broadly elliptic to circular, with an obscure notch and marginal scar; surface cellular-reticulate or smooth; yellow to brown. *Solanum* and *Physalis* seeds are often confusingly similar, but usually they can be distinguished from each other. *Solanum* seeds vary more in size (1.5–5 mm. long), shape, and surface, whereas *Physalis* is about 1.5–2 mm. long and is consistently **shiny, cellular-reticulate,** and broadly oval.

Fig. 265. *Solanum verbascifolium* X 7.5

Fig. 266. *Solanum dulcamara* X 7.5

Fig. 267. *Physalis heterophylla* X 7.5

SCROPHULARIACEAE Figwort Family

Ordinarily minute to small; diverse in shape; surface often soft or **frothy with cellular reticulations**; embryo linear to spatulate in fleshy endosperm.

VERBASCUM Mullein (pl. 252)

Cylindric and truncate at both ends, with **wavy lengthwise ridges** and a dark spot at each end; resembling caterpillar droppings; about 1 mm. long or less and about one-third as thick.

Fig. 268. *Verbascum thapsus* X 25

LINARIA Butter-and-eggs, Toadflax (pls. 253, 254)

Seeds **diverse**; in *L. vulgaris* they are flat, circular-winged, dark brown, about 2 mm. across, with pimple-like papillae over the central area and radial striations evident along the margin; *L. canadensis* seeds are dark grayish, oblong-quadrandular and somewhat wedge-shaped (both ends truncate), about 0.5 mm. long.

Fig. 269. *Linaria canadensis* X 30

PENSTEMON Penstemon (pl. 255)

Irregularly angled, generally sharply; margins **often narrowly winged**; surface finely cellular; black or brownish; 1–2 mm. long.

Fig. 270. *Penstemon barbatus* X 9

Fig. 271. *Penstemon breviflorus* X 15

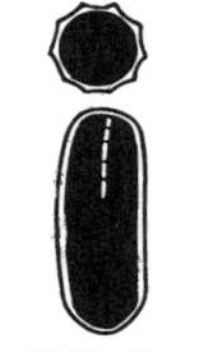

Fig. 272. *Bacopa acuminata* X 30

BACOPA Waterhyssop (pl. 664)

Oblong, and quadrangular or cylindric; truncate at both ends; **minute, about 0.5 mm.** long; surface **cellular-reticulate in lengthwise rows**, 2 or 3 rows per side; cream (*B. monniera, rotundifolia*) or black (*B. acuminata*).

Fig. 273. *Veronica anagallis-aquatica* X 25

VERONICA Speedwell (pl. 819)

Generally oval to **circular, and flat** or flattish, often slightly concave-convex; the concave side has a small opening toward the center and a dark dot near the margin; yellow to brown; usually **semitransparent**; 0.5–1 mm. across. *V. virginica* is oblong-cylindric and arched.

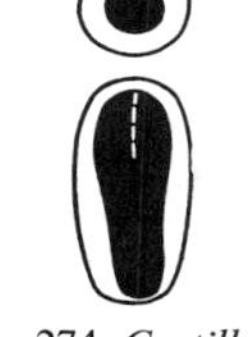

Fig. 274. *Castilleja indivisa* X 23

CASTILLEJA Paintbrush (pl. 256)

Irregularly shaped, sometimes variously angled, **in a mesh-bag network of coarse cells;** light brown to black; about 2 mm. long; frequently the seed is loose inside the enclosing network.

BIGNONIACEAE Bignonia Family

Seeds flat; **bilaterally winged;** embryo with large cotyledons which invest the stalk; endosperm none.

TECOMA Trumpetvine (pl. 793)

Wings thin, transparent, with wavy lines radiating from the center; 2–2.5 cm. long; body area circular-ovate, brownish.

CATALPA Catalpa (pl. 794)

Wings silky-papery, whitish, with obscure lines radiating from the central area; 2.5–3 cm. long.

PEDALIACEAE Pedalia Family

Fig. 275. *Sesamum indicum* X 4.5

SESAMUM Sesame (pl. 527)

Flattish-ovate, with dark base and **faint marginal lines** and an equally faint central line on one face; white to light brown; 3 X 2 X 1 mm.

PLANTAGINACEAE Plantain Family

PLANTAGO Plantain (pls. 257–259)

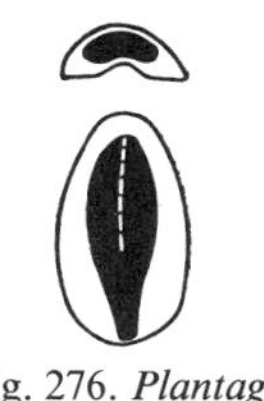

Fig. 276. *Plantago rhodosperma* X 7.5

In most species, the seed is **hollowed out** like a boat, but it is plano-convex and elliptic in *P. elongata, eriopoda, macrocarpa, maritima,* and *tweedyi*; irregularly angled in *P. major* and *rugelii*; 1.5–4 mm. long.

RUBIACEAE Madder Family

Seeds (some of them nutlets) diverse: globose, ovoid, flat and lanceolate, or other shapes; minute to large; embryo linear to spatulate in fleshy endosperm.

CEPHALANTHUS Buttonbush (pl. 795)

Fig. 277. *Cephalanthus occidentalis* X 5

Cylindric-lanceolate, like miniature wooden pegs, with a dark cap on the broad end; **narrow lengthwise groove** on opposite sides of the 2-seeded nutlet (as evident in cross section); about 5 mm. long.

MITCHELLA Partridgeberry (pl. 820)

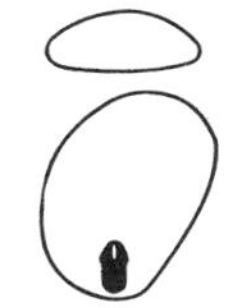

Fig. 278. *Mitchella repens* X 6.5

Flattish oval-oblong; with a **small hole or protrusion** on one face and with fibers of the wall concentric about this point; whitish; 3 X 2 X 1 mm.

DIODIA Buttonweed (pl. 260)

Fig. 279. *Diodia teres* X 6

Oval-oblong; *D. teres* is **plano-convex**, with a thick end topped by 3 apical bracts, and the other end thin; its flattened face is shallowly indented, and the rounded back is bristly-hairy. In *D. virginiana,* 2 facing seeds commonly remain attached to each other in a **compressed oval-oblong** body with about 10 lengthwise, broadly rounded ribs separated by grooves. The seed of *D. virginiana* is 6–7 mm. long, whereas that of *D. teres* is about half as long.

GALIUM Bedstraw (pls. 821–823)

Globose or subglobose, hollow-centered or hollowed out on one side; bristly, tuberculate, or smooth; black or dark brown; 1–3 mm. across.

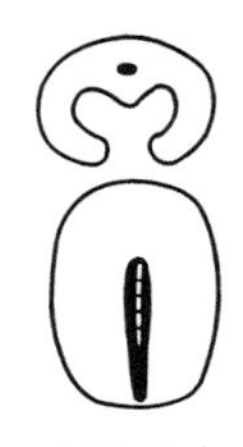

Fig. 280. *Galium bermudense* X 6

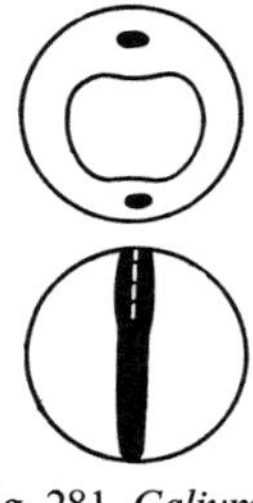

Fig. 281. *Galium trifidum* X 13

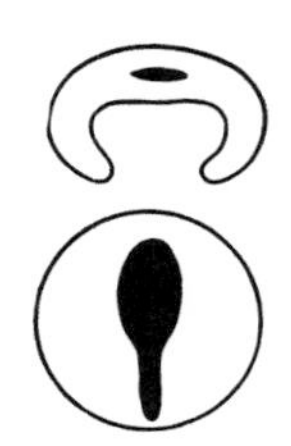

Fig. 282. *Galium latifolium* X 6

CAPRIFOLIACEAE Honeysuckle Family

Seeds (some of them nutlets) diverse in shape and size, mainly **woody-walled; embryo small or medium-sized,** in fleshy endosperm.

SAMBUCUS Elderberry (pl. 796)

Elliptic-ovate and compressed rounded-triangular; generally **slightly arched** inward; with **crosswise wrinkles;** brown, dull; wall woody; 2–3 mm. long.

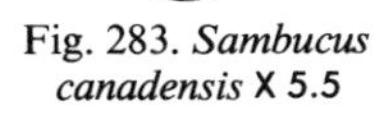

Fig. 283. *Sambucus canadensis* X 5.5

SYMPHORICARPOS Snowberry, Coralberry (pls. 797, 798)

Elliptic plano-convex, with rounded edges and a small, porelike opening at one end; wall **tough, woody, with obscure lengthwise fibers;** white to light brown; 3–4 mm. long.

Fig. 284. *Symphoricarpos orbiculatus* X 6

LONICERA Honeysuckle (pls. 799, 800)

Oval to elliptic, flattish or compressed, with a notch or point at one end; often irregularly ridged and warped; finely to coarsely cellular; yellow to dark brown; wall crustose to woody; 2–5 mm. long.

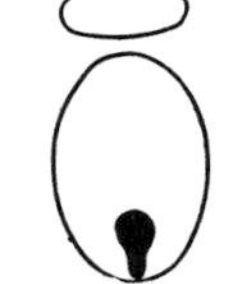

Fig. 285. *Lonicera dioica* X 4.5

VIBURNUM Viburnum (pls. 801–803)

Diverse; flat or flattish and ovate, elliptic, or circular, to ovoid or globose; with **rounded lengthwise ridges and grooves;** wall hard but generally rather thin; 4–10 mm. long.

Fig. 286. *Viburnum dentatum* X 4.5

CUCURBITACEAE Melon Family

CITRULLUS Watermelon (pl. 528)

Flat, oval-ovate, with a **marginal groove on each side near the base;** white, black, or white with black margins; 10–15 mm. long.

CAMPANULACEAE Campanula Family

SPECULARIA Venus Looking-glass

Flattish or compressed, oval to elliptic, smooth, dull to glossy; **minute,** generally less than 0.5 mm. long; *S. biflora* is glossy light brown, and *S. perfoliata* is broadly oval, thick, purplish brown, and dull.

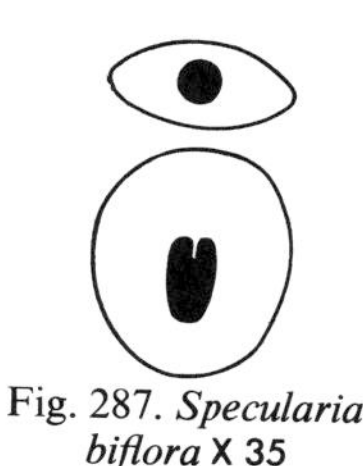

Fig. 287. *Specularia biflora* X 35

COMPOSITAE Daisy Family

Seeds (achenes) diverse; **many of them oblong or elongate, with or without a terminal ring (pappus) of hairs or scales;** characteristics of the pappus are often useful for identification; **embryo with well-developed cotyledons and a broad stalk having a truncate top;** endosperm none.

VERNONIA Ironweed (pl. 665)

Body **oblong, coarsely ridged,** light brown, 3–4 mm. long; pappus generally purple-tinged and longer than the body; the hairs **short-plumose** with very fine secondary hairs; base of pappus encircled by a series of semitransparent, **flat, hairlike scales.**

LIATRIS Gayfeather (pl. 261)

Similar to *Vernonia* in **plumose pappus,** which is purple-tinged in some species; no pappus scales; **body generally black, hairy, narrowed toward base;** about 5 mm. long.

EUPATORIUM Boneset (pl. 666)

Body **oblong, 4- or 5- sided, black,** tapering toward base and often white-tipped; length variable, from 1.5 mm. (as in *E. coelestinum, mikanioides, urticaefolium*) to 5 mm.; **pappus of delicate, stiff white hairs, which are minutely plumose.**

CHRYSOPSIS Goldaster (pl. 263)

Body flat, ovate to elliptic-oblong; hairy; light-colored to brownish (*C. graminifolia*); 2–3 mm. long; **pappus double, hairs of inner ring brownish,** stiff, minutely plumose, **outer ring composed of short, hairlike scales.**

GRINDELIA Gumweed, Rosinweed (pl. 262)

No pappus; body oblong, compressed or flat; truncate at top, narrowed toward base; often arched sideways, with one edge straight or concave and the other convex; surface marked by lengthwise lines or irregular ridges; straw-colored or brown; length 2 mm. (*G. squarrosa*) to 5 mm.

GUTIERREZIA Snakeweed (pl. 264)

Body **oblong-ellipsoid; small** (1–2 mm. long); dark, clothed with whitish hairs; **pappus a ring of scales.**

HETEROTHECA Heterotheca (pl. 265)

Very similar to *Chrysopsis* in having a double pappus with brownish pappus bristles, surrounded by hairlike scales, and in its flat, oblong-attenuate, light-colored, hairy body. The ray-flower achenes of *Heterotheca* are distinct in being **smooth, compressed-triangular, and without pappus.**

SOLIDAGO Goldenrod (pls. 266, 267)
ASTER Aster (pls. 667, 824)
ERIGERON Fleabane (pl. 268)

Seeds of these extensive genera overlap considerably in size, shape, and other characteristics. The seed body is small, ordinarily 1–3 mm. long, oblong-cylindric to compressed oblong-elliptic, and with a tapering base, hairy or smooth; pappus is a ring of white, minutely plumose bristles, which are often wavy near their base. *Solidago* seeds average smaller in their body (generally 1–2 mm. long) than *Aster*, and are more consistently hairy and oblong-cylindric; *Aster* seeds are rather distinct from *Solidago* in commonly being **compressed** and in having **lengthwise ridges** on the fairly numerous species which have smooth or nearly smooth seeds. *Erigeron* seeds, except for species like *E. salsuginosus,* which are comparatively large (body 2.5 mm. long) and coarse, can usually be distinguished from those of *Aster* and *Solidago* by the **delicate pappus on a smaller body** (1–1.5 mm. long), which is **flattish, yellow to brown.** *Erigeron* seeds vary in pubescence from sparsely short-hairy to densely hairy.

IVA Marshelder, Sumpweed (pls. 668, 669)

Flattish or compressed, obovate to broadly oblanceolate and **truncate** at the top; black or brown; dull and often scurfy; **wall interior white;** 2–5 mm. long.

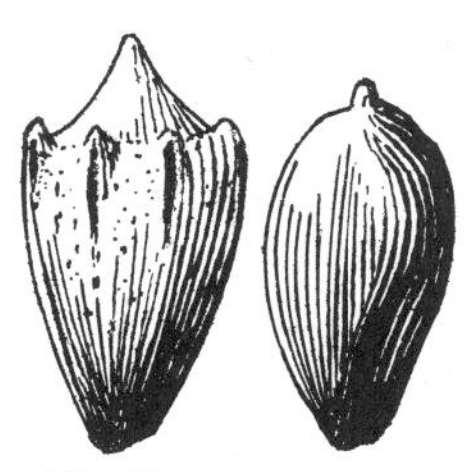

Fig. 288. *Ambrosia artemisiaefolia*

AMBROSIA Ragweed (pls. 269–272)

Obovoid to oblanceolate, generally with spines near the top, the spines often buttressed by ridges; 2–10 mm. long; usually grayish, but on worn seeds of the common species, *A. artemisiaefolia*, the gray coat gives way to a shiny brown one, which in turn overlies successive black and white layers.

RUDBECKIA Coneflower, Black-eyed-Susan (pls. 273, 274)

Oblong to linear and generally 4-sided (cylindric or compressed-cylindric in *R. amplexicaule*, which is distinct in its fine cross wrinkles on a black coat); upper end truncate; sides usually with fine lengthwise lines; short pappus scales present or absent; gray-brown to black; 2–8 mm. long.

RATIBIDA Prairie-coneflower (pl. 275)

Flat to compressed, oblong, **sharply truncate at top** and rounded at base; obscure lengthwise lines on black or grayish black surface; with or without a pair of pappus scales; 2–3 mm. long. *R. peduncularis* seeds are margined with a fringe of flat, finger-like projections.

HELIANTHUS Sunflower (pls. 276–278)

Compressed or flattish, oblong; top truncate and base narrowed; generally grayish, with **dark lengthwise streaks** and mottlings; often pubescent, especially near the top, with stiff, upward-directed hairs; attachment area shiny; mostly about 5 mm. long, but in the cultivated form of *H. annuus* the length may exceed 1 cm.

ENCELIA Encelia (pls. 279, 280)

Flat, broadly-oblanceolate; a dense fringe of long white hairs on the margins and with or without hairs elsewhere; 5–10 mm. long; *E. farinosa* is emarginate at the top.

VERBESINA Crownbeard (pls. 281–283)

Flat to compressed and oval to oblong; most species are wing-margined and have a **deep apical notch** which is bordered by 2 obscure awns, but *V. occidentalis* is nearly oblong, wingless, and has 2 long, hornlike awns; about 5 mm. long.

MADIA Tarweed (pls. 284, 285)

Flattish-oblanceolate and often arched sideways; with fine lengthwise lines; black or grayish; 3–4 mm. long.

HEMIZONIA Tarweed (pls. 286, 287)

Rounded-triangular and generally with its broad face arched inward; style base knoblike; black or grayish; 2–3 mm. long.

COREOPSIS Coreopsis (pl. 288)

Flat or flattish, and either circular, oval, or oblong; often arched inward like a curved potato chip; wing-margined in certain species (*C. gladiata* has finger-like marginal projections) and frequently **tuberculate** on one or both faces; 1.5–6 mm. long.

BIDENS Sticktight (pls. 670, 671)

Linear and several-sided, to flat-oblong or flat-oblanceolate; **2 to 4 awns,** which are generally **barbed**, extend upward from the top, and some species have barbs on the margins of the seed body; brown to black; entire seed, including awns, 0.5–2 cm. long.

HELENIUM Sneezeweed (pls. 290, 291)

Body obconic or oblong and several-sided; crowned with about **6 semitransparent, broad, tapering pappus scales**, which are frequently tipped by awns; entire seed 2–7 mm. long. *H. tenuifolium* seeds are obconic and have long hairs extending upward from the base.

GAILLARDIA Gaillardia (pl. 289)

Body obconic; densely coated with stiff hairs; **crowned by 6 or more long-awned pappus scales;** brown; entire seed 5–7 mm. long.

ANTHEMIS Mayweed (pl. 292)

Usually **oblong and 4-sided**; crowned by a cup in *A. arvensis* and *tinctoria; A. cotula* is club-shaped, with about **8 lengthwise, rather knobby ridges**, and has a rounded top; 1.5–2 mm. long.

ACHILLEA Yarrow (pl. 293)

Flat-oblong; truncate at the top and narrowed to a rounded base; the **thin margins whitish** and the body dark; pappus none; about 2 mm. long.

CHRYSANTHEMUM Oxeye Daisy, Chrysanthemum (pl. 294)

Cylindric; with numerous **lengthwise ridges**; *C. coccineum* and *uliginosum* are cup-shaped at the top; *C. leucanthemum* and *maximum* are either rounded or truncate at the top, have a prominent, knoblike style base, and a body striped with light-colored ridges alternating with dark lines; about 2–3 mm. long.

CIRSIUM Thistle (pls. 295, 296)

Body compressed, lanceolate-oblong; **top cup-shaped** and usually **tilted slightly sideways**; body narrowed to a rounded base and marked by obscure lengthwise lines; gray to dark brown, smooth; 3–6 mm. long; pappus composed of many long, plumose bristles, which are united at the base into a readily detached ring.

CENTAUREA Starthistle, Centaurea (pls. 297-299)

Body cylindric or compressed-cylindric, elliptic-obovate or elliptic-oblong; **top truncate; base asymmetric and notched,** or in some species, the base is narrowed and arched like a hook; pappus white or tawny, persistent or deciduous, composed of stiff, bristle-like scales, making the seed look **like a shaving brush**; body smooth, gray to black; 2.5–5 mm. long.

SILYBUM Milk Thistle (pl. 300)

Body compressed oblong-lanceolate; **top cup-shaped, tilted slightly sideways,** and enclosing a large, **knoblike style base**; body black with white collar; 7 X 3 X 2 mm.; pappus deciduous.

CICHORIUM Chicory (pl. 301)

Slightly compressed and arched-oblong; **top truncate and fringed with a ring of stubby scales;** body narrowed into a rounded base and **minutely cross-wrinkled**; brown or mottled; 2 X 1 X 0.75 mm.

SERINEA Serinea (pl. 302)

Elliptic-cylindric; narrowed toward base; marked with about **12 lengthwise, finely cross-wrinkled ridges;** reddish brown or purplish brown; about 1 mm. long and less than half as wide; pappus none.

TARAXACUM Dandelion (pl. 303)

Body flattish-oblanceolate and somewhat 4-angled; narrowed toward both ends; **tubercles or teeth** near the upper end of the lengthwise ridges; yellow or straw-colored; 3–4 mm. long; the pappus delicate, white, and appearing **like an inverted umbrella** on a long handle.

SONCHUS Sowthistle (pl. 304)

Body **flat, and elliptic or lanceolate-elliptic**; similar to *Lactuca* in shape, ridging, and minute wrinkling, but **smaller**, about 3 mm. long, and **more obtuse** at ends; brown or reddish brown; pappus bristles long, soft, delicate, deciduous.

LACTUCA Wild-lettuce (pls. 305, 306)

Body **flat, elliptic to lanceolate**; pointed near both ends, though in *L. spicata* the apical end broadens again into a terminal disc; most species have about **8 lengthwise ridges**, which are finely cross-marked, but *L. canadensis* has only 1 conspicuous ridge and is finely cross-wrinkled over the whole surface, while *L. spicata* has 2 ridges and is similarly cross-marked; pappus delicate, deciduous; body grayish brown to black, except in *L. sativa*, which is cream-colored; body 4–5 mm. long.

HIERACIUM Hawkweed (pl. 307)

Body **cylindric**; top truncate; body narrowed toward base and marked by about **10 distinct lengthwise ridges**, which in some species are finely cross-marked; **dark reddish brown to black;** 2–4 mm. long; pappus bristles stiff, delicate, in a single series.

Selected Bibliography

Family Characters

Investigations in Seed Classification by Family Characteristics. Duane Isely. Iowa Agr. Exp. Sta. Res. Bull. 351. 1947.

Seed Characters of Selected Plant Families. David S. McClure. Iowa State Coll. Jour. Sci. 31:649–682. 1957.

Weed Seeds

Seed Drawings. Pls. 1–15. U. S. Dept. Agr., Office of Information, Division of Photography. [Fifteen sheets of photograph reproductions of excellent seed drawings, twenty-four species per page.]

Unkrautsamen [Weed Seeds]. Emil Korsmo. Oslo, Grondahl & Sons, 1935.

Woody Plants

Graines et plantules des conifères. R. Hickel. Macon, France, 1911, 1914.

Trees: The Yearbook of Agriculture, 1949. Washington, U. S. Dept. Agr., 1949.

Woody-Plant Seed Manual. U. S. Dept. Agr. Misc. Publ. 654. 1948.

Comparative Morphology

The Comparative Internal Morphology of Seeds. Alexander C. Martin. Am. Midl. Nat. 36:513–660. 1946.

Seed Testing

Manual for Testing Agricultural and Vegetable Seeds. U. S. Dept. Agr. Handbook 30. 1952.

Families and Genera

ATRIPLEX

An Illustrated Flora of the Pacific States. Leroy Abrams. Vol. 2. Stanford, Calif., Stanford Univ. Press, 1944.

BETULA

A Seed Key for Five Northeastern Birches. Frank E. Cunningham. Jour. For. 55:844–845. 1957.

BRASSICA

Distinguishing Species of *Brassica* by Their Seeds. Albina F. Musil. U. S. Dept. Agr. Misc. Publ. 643. 1948.

Seeds and Seedlings of the Genus *Brassica.* Jean M. McGugan. Canad. Jour. Res., Sec. C, 26:520–587. 1948.

Seeds of Commercial Species of *Brassica.* Albina F. Musil. Proc. Assn. Offic. Seed Anal. 34:132–138. 1942.

CAREX

North American Cariccae. K. K. MacKenzie. New York, N. Y. Bot. Gard., 1941.

CHENOPODIUM

Seeds of the North American Species of *Chenopodium.* Sister M. Angelita Conley. Master's Thesis, Univ. Notre Dame. 1938.

COMPOSITAE

Akenes of Some Compositae. Anita Mary Blake. N. Dak. Agr. Coll. Bull. 218. 1928.

CRUCIFERAE

Seeds of the Cruciferae of Northeastern North America. Margaret R. Murley. Am. Midl. Nat. 46:1–81. 1951.

CUSCUTA

Seeds of Dodder Occurring with Crop Seeds. Albina F. Musil. Washington, U. S. Dept. Agr., Bur. Pl. Ind., Soils and Agr. Eng., 1944.

CYPERUS

A Revision of the Subgenus *Eucyperus* Found in the United States. Sister M. Vincent de Paul McGivney. Cath. Univ. Am. Publ., Biol. Ser., 26. 1928.

ELEOCHARIS

A Flora of the Marshes of California. Herbert L. Mason. Berkeley and Los Angeles, Univ. Calif. Press, 1957. [Cited, after first reference, as Mason.]

Monographic Studies in the Genus *Eleocharis.* H. K. Svenson. Rhodora 31:121–135, 152–163, 167–191, 199–219, 224–242; 34:193–203, 215–227; 36:377–389; 39:210–231, 236–273; 41:1–19, 43–77, 90–110. 1929, 1932, 1934, 1937, 1939.

ERODIUM

Alfileria (Filaree) Seed. William A. Dayton. Rhodora 39:323–325. 1937.

EUPHORBIA

Euphorbia, Subgenus *Chamaesyce,* in Canada and the United States, Exclusive of Southern Florida. Louis Cutter Wheeler. Rhodora 43:97–154, 168–205, 223–286. 1941.

GERANIACEAE

A Seed Key to Fourteen Species of Geraniaceae. Margaret R. Murley. Proc. Iowa Acad. Sci. 51:241–246. 1944.

GRAMINEAE

Manual of the Grasses of the United States. A. S. Hitchcock. 2d ed., rev. by Agnes Chase. U. S. Dept. Agr. Misc. Publ. 200. 1950.

LEGUMINOSAE

Anatomical Characters of the Seeds of Leguminosae, Chiefly Genera of Gray's Manual. L. H. Pammel. Trans. Acad. Sci. St. Louis 9:91–263. 1899.

POLYGONUM

Identifying Polygonum Seeds. Alexander C. Martin. Jour. Wildl. Mgmt. 18:514–520. 1954.

POTAMOGETON

Identifying Pondweed Seeds Eaten by Ducks. Alexander C. Martin. Jour. Wildl. Mgmt. 15:253–258. 1951.

RHYNCHOSPORA

Rhynchospora, Section *Eurhynchospora,* in Canada, the United States, and the West Indies. Shirley Gale. Rhodora 46:89–134, 159–197, 207–249, 255–278. 1944.

UMBELLIFERAE

Umbelliferae in Iowa, with Seed Keys. Margaret R. Murley. Iowa State Coll. Jour. Sci. 20:349–364. 1946.

Index

This index is based primarily on genus and family names, both scientific and common. Photographic plates 308–510 are indicated by italic numbers because they are duplicates of plates in the preceding series, 1–307.